"Dr. Hirsch's book is a sensory map that teaches you how to, literally, follow your nose back to the playful, lusty, satisfying sex you once had."

Mary Hogan, magazine writer, *Family Circle, Parenting, First For Women, Fitness*

"*Scentsational Sex* offers a provocative and fascinating look at what really turns us on. Mom lied. The quickest way to a man's heart—and other parts—is through his nose."

Mary-Ellen Banashek, senior writer, *New Woman*

"Dr. Hirsch's *Scentsational Sex* is a well-crafted blend of fascinating facts and practical advice. Ultimately it's a satisfying read, guaranteed to bring the nose back into the bedroom."

Lois B. Morris, "Mood News" columnist, *Allure* magazine

"Nobody is better qualified than Dr. Hirsch to write the authoritative guide on making good use of our most neglected sexual organ, the nose. He expertly sorts through myths and hearsay on pheromones and aphrodisiacs, and comes up with hard scientific data on what scents can actually improve your sex life. His advice is fun to read—and even more fun to put to use."

Zachary Veilleux, *Sex & Health* newsletter for men

"Fascinating! Useful! Try it. You'll like it, too . . . really like it!!!"

Martin Edelston, editor, *Bottom Line Health*

"The mystery of scent and its link to sensuality and sexuality is unlocked in this highly readable work by the leading U.S. expert on smell and taste research. Dr. Hirsch has another winner on his hands with this clear, laymen-friendly (yet scientifically sound) book about understanding 'scent secrets' to sexuality."

Peggy Noonan, writer for *First for Women, Family Circle, Men's Health* and *Longevity*

Scentsational
SEX

*The Secret to Using
Aroma for Arousal*

ALAN R. HIRSCH, M.D., F.A.C.P.

E L E M E N T

Boston, Massachusetts • Shaftesbury, Dorset
Melbourne, Victoria

Published in Great Britain in 1998 by
Element Books Limited
Shaftesbury, Dorset SP7 8BP

First published in the USA in 1998 by
Element Books, Inc.
160 North Washington Street,
Boston, Massachusetts 02114

Published in Australia in 1998 by
Element Books Limited for
Penguin Books Australia Limited
487 Maroondah Highway, Ringwood, Victoria 3134

British Library Cataloguing-in-Publication Data available

Printed and bound in the United States by Edwards Brothers, Inc.

ISBN 1-86204-240-3

Contents

To my wife, Debra, and our two children, Marissa and Jack, and to the individuals who participated in our studies and helped us investigate the important relationship between odors and sexuality

Acknowledgments

MANY PEOPLE supported my efforts to produce this manuscript, not the least of whom are the hardworking staff at the Smell and Taste Treatment and Research Foundation in Chicago, Illinois. Daily, they put forth their efforts in advancing research on olfaction, and their assistance to me is invaluable. Special thanks are in order for Denise Fahey, Charlene Bermele, Deborah Zagorski, Marianne Schroder, and Jason Gruss.

I extend special thanks to Roberta Scimone at Element Books for enthusiastically embracing the idea for this book, and to Virginia McCullough for helping me produce it.

Many of my colleagues and friends have supported my research endeavors. To name a few would mean leaving out too many, so I offer my appreciation of them as a group, with special mention to Dr. Jan Fawcett, Chairman, Department of Psychiatry, Rush Presbyterian St. Luke's Medical Center. I also want to thank Dr. Sally Freels at the University of Illinois School of Public Health for assistance with statistical analysis. As always, Stan Block has been a true friend.

My family has been patient and indulgent as I pursued this and other work. I am especially grateful to my wife, Debra, for her tolerance and understanding.

Foreword

FINALLY, THERE is scientific support of what we sex goddesses and gods have instinctively known for eons: the *nose* has a mind of it's own and the right aroma can be a magical thing for lovers!

Sensual sorceresses through the ages have regularly utilized tangy passion scents to entice and seduce, as well as secure their relationships with their men. Cleopatra came to Rome in a boat with sails saturated in jasmine to lure Mark Anthony to her floating boudoir and Empress Josephine used her natural odors along with violets to keep her conqueror Napoleon wild with lust for her. Just as women in history have appealed to the primal instincts of the human male's proboscis, the modern woman and her willing partner can have much fun with the suggestions in this wonderful book, *Scentsational Sex: The Secret to Using Aroma for Arousal.*

I first came across Dr. Alan Hirsch's work as Editor-in-Chief of *Playgirl*. His famous study results on food odors and sexual arousal in men had become such a significant part of sex research that his work was already filtering through to us in books and videos we received for review, yet it didn't always come through with precise attribution. For instance, it was Dr. Hirsch's finding that black licorice, in combination with donuts, increased penile blood flow by 31.5 percent, yet it was circulated through the sex world as if it was part of some mystical folk lore. That's when I began to walk around with black licorice in my pocketbook, simply because, as *Playgirl's* editor, I never knew when I would come across a male who required some support in increasing penile blood flow. (Erections were often a requirement for *Playgirl* models, who occasionally got shy in front of the camera.) In hindsight, we should have doused the set in lavender and served the male models pumpkin pie and donuts for the best, most consistent results. Oh well. You live and learn.

Our paths crossed again while I was Editor-in-Chief of *Single Living Magazine*. While working on a story on his first book, *Dr. Hirsch's Guide To Scentsational Weight Loss*, my staff and I decided to test out his findings. We walked around sniffing a test tube vial with a tiny, green apple scented cotton ball in it, as many times a day as we could bear it. We were like green apple junkies. We all dropped around five pounds.

I think my most pleasant experimentation with Dr. Hirsch's work was while researching my most recent book, *How To Seduce A Man and Keep Him Seduced*. For that project, my significant other (a sex therapist) agreed to some *scent*sational sex and surrendered himself as my guinea pig. I used potent scents in strategic locations: vanilla under his nose, strawberry under my nose (so he could sniff it when we kissed), pumpkin pie scent over the light bulb and lavender all over the sheets.

I don't know which scent worked best (we lost our desire, and ability, to keep it scientific after a while) but boy did we have fun— and you can too if you make *scent*sational sex a fun game, rather than a science project!

I might add, that it's thanks to Dr. Alan Hirsch, and his break-through discovery on the types of food odors that get a guy's blood flowing in all the right places, my sweetheart and I also enjoyed a very special "picnic" to further "test out" his findings.

The spread included pumpkin pie, hot cinnamon rolls, licorice and fresh strawberries. I fed my beloved bite-sized pieces of the recommended foods, letting each morsel linger beneath his nose for a long time before popping it in his mouth. I kept my hands to myself, but it wasn't long before he had a hard time keeping his hands, and other parts, to himself. I believe there was a point at which my body became the picnic buffet. Again, I didn't measure which food scents had the greatest impact on penile blood flow, I simply enjoyed the results of that increased circulation!

As Dr. Hirsch points out in this book, the triple combination of odor, food and sex may date back to the time of the earliest humans. Perhaps they were lured by the scent of food, joined a group for dinner and then had the opportunity of having sex for dessert. It seems that modern couples can recreate these early rituals (no need to invite the neighbors, though!). Odors can be utilized as attractants and stimulants and food can be part of erotic playfulness.

You don't have to be a scientist to figure out how to use simple aromas and aphrodisiac foods in your love play. There are so many types of essential oils, massage lotions and scented candles to choose from. And finger food is a delicious part of good sex (For example: just slowly bite into that strawberry, sensually rub on your lips, then outline your beloved's lips with the deliciously dripping fruit. Just one little kiss and a sniff can lead to a night of ecstasy).

Beyond that, you can weave your special aroma into your beloved's heart (and very being) through association with a certain scent. A little tangerine, jasmine or ylang-ylang behind the ears, or under the arms, can go a long way. And you can keep the sensual flames burning by carrying over your special sex scent from your body to the boudoir; then dab a little in his or her briefcase, so that your beloved can take the essence of your love everywhere!

This book is chock full of insights and evidence that can lead you to successful *scent*sational sex. May you take advantage of the opportunity to learn how to use aroma to attract, heal, play and love!

LAURIE SUE BROCKWAY
author of *How to Seduce a Man and Keep Him Seduced*

Preface

ALTHOUGH ODORS can be used systematically to enhance many areas of life, this book concentrates on odors and the areas of sensuality and sexuality. My previous book, *Dr. Hirsch's Guide to Scentsational Weight Loss*, addressed the specific problem of obesity and explained the way in which our sense of smell can become an ally when we're trying to treat weight problems or are attempting to maintain a healthful weight. This guide to scentsational sex has even wider implications, with the potential to benefit every sexually active adult, because it offers information about what was once considered suitable for the realm of folklore and literature, but probably not scientifically important.

You don't need to be a scientist to benefit from this book, and I expect most readers are interested in the practical applications of what we have learned from our studies on scent and sexuality. It may surprise you to know that certain odors can actually increase sexual arousal in men and women, a research finding that has implications for the medical treatment of impotence or decreased sexual desire, for example. And, medical applications aside, a vast array of odors can add to our enjoyment of sex.

In other words, you needn't have sexual "problems" to have a good time with the issues explored in this book. I hope you will use the information in order to have more fun. Certainly, I'll take you through the facts of sensuality and odors. You'll learn about the way *olfaction*—the term for the sense of smell—works, and this understanding will help you use many different odors to add pleasure to your sexual experiences.

I wrote this book to give the general public a glimpse of what we are able to demonstrate scientifically about our sense of smell and how it influences our sexual attractions and expression. I've also added some information that comes from popular culture and aroma-therapy that hasn't yet been scientifically proven, although some of this information may be verified in the near future.

Attraction, love, and sex are—and always have been—complex and mysterious, and no final word on the subject yet exists. So, use this book to broaden your understanding of yourself and your partner(s) and enjoy your sensuous sense of smell.

ALAN R. HIRSCH, M.D., F.A.C.P.
CHICAGO, ILLINOIS

1

The Sexy, Sensuous Nose

LONG, LINGERING KISSES, the sensual texture of cool silk sheets, the sound of romantic music, and the sight of our lover's flesh—all are classic images of romance and sex. We see these fantasies acted out on television and in movies, and they are part of our shared culture. But something is missing from this fantasy, and it could be the most important element. Where are the fragrances we associate with romance? Where are the odors that add so much to the mood we enjoy?

The scents we associate with sexuality and romance are not easy to describe, much less depict, and sometimes we even forget their importance in our sexual lives. Yet, we're in the midst of an odor revolution, so to speak, one that is bringing a growing appreciation of our sense of smell and its power to enrich all areas of our lives, including our sexuality. Never have so many people been so conscious of the odors they use to enhance their surroundings and create or promote a variety of moods.

We live in an opulent world of odors, with abundant opportunities to add pleasant smells to our environment. The booming commercial arena of odor research is eager to learn all about our odor preferences and happily accommodates us as we seek to try new fragrances. If we like odors that make us think about nature, we can buy special soaps and shampoos whose added odors mimic the smells of the

1

outdoors. We call these odors "natural," even though we know the odors are artificially produced. We smooth odorized lotions on our bodies, put special fragrances in our closets to alter the smell of our clothing, and millions of us use scented candles throughout our homes.

In truth, humans have always used smells to alter their environment, but in the past we chose particular scents based not only on what was appealing, but also on the cultural myths with which they were linked. Some odors have specific uses, such as those used in religious rituals and other special smells that remind us of holiday memories and their cultural roots. Virtually every society uses odors in its marriage rituals and other rites of passage.

Some scents are used to add to an experience generally associated with a different sensory stimulation. We usually link massage with our sense of touch, yet many massage oils, for example, are infused with odors intended to relax us after a strenuous workout or soothe us when stress has us frazzled and tense. Other such oils are odorized with fragrances meant to promote a romantic mood. Such a wide variety of scents is available that it is easy for us to experiment and find the odors that add to our total enjoyment of life, including our sexuality.

Science Meets Folklore

Virtually all cultures have used odors in their sexual rituals. By that I mean that all available anthropological evidence shows us that in one way or another all the senses have been incorporated into sexual life. But, for most of human history, we have had no way to prove in a laboratory that different kinds of sensory stimulation had the power to alter mood and change behavior.

Today, thanks to technology and science, we have more sophisticated knowledge about our senses. We know that adding certain sounds, colors, and odors to an environment can actually change brain wave activity. Our other senses are similarly linked to mood. For example, if we play certain kinds of music, we'll feel more relaxed, and our brain wave pattern will be different from that induced by the booming sound of heavy metal rock.

Our sense of smell is a special partner in creating changes in brain wave patterns and altering our moods.[1] If we introduce the odor of lavender in the environment of a person experiencing anxiety, for example, an EEG (electroencephalogram) will show increased alpha brain wave activity because the odor alone has caused the person to experience less anxiety. Alpha brain waves, which occur in the back of the brain, are those associated with a relaxed mental state. The odor of jasmine can alter beta wave activity in the brain. Beta waves, which occur in the front of the brain, are associated with an awake and alert mental state.

Recent research has confirmed what most people instinctively know: odors can soothe frazzled nerves, promote sleep, wake us up, lift our spirits, and help us learn and work more efficiently.[2] Individual odor preferences and cultural uses of particular scents have a sound scientific basis. We can use this information and consciously improve our quality of life in every area. And there's every reason to use scents to enhance our sexual experiences, which ironically, is why our sense of smell has not always had an elevated status.

Odors versus Intellect, or, Can Good Smells Lead Us Astray?

When cultural anthropologists, who in the early years of the field came from Western cultures, discovered evidence of the use of odors in the sexual life of ancient—and not so ancient—peoples, they made an unfortunate link that caused some people to view the sense of smell with suspicion. Briefly, these anthropologists showed their cultural bias by declaring that it was primarily the so-called "primitive" societies who relied on their sense of smell in their rituals, sexual and otherwise.

Intuitively, however, we know that our sense of smell is very powerful, and we depend on it, enjoy it, and miss it when injury, illness, or the normal aging process diminishes our ability to inhale pleasant odors in our environment. Of course, poets throughout the ages have spoken eloquently about the sense of smell, particularly when describing their feelings about the natural world and sexual desire. The nineteenth-century French poet Baudelaire was particularly expressive about the smells of eroticism. This is what he wrote about his lover:

> In bed her heavy resilient hair
> —a living censer, like a sachet—
> released its animal perfume,
> and from discarded underclothes
> still fervent with her sacred body's
> form, there rose a scent of fur.

This entire passage concerns his lover's special scent, which even when he isn't with her, can be recalled in splendid detail. Jumping ahead a couple of centuries, Sandra Brown, a best-selling romance novelist, makes odor the central theme of this description of a sexual encounter of two characters in her novel *French Silk*: "'This is dangerous as hell, but. . . . Ah, God, I can smell you.' He leaned forward and burrowed his face in the cleft of her thighs, nuzzling, gnawing, kissing madly through the giving fabric of her dress. 'Too bad you can't bottle this.'"

Different words, same basic response to the odors of sexuality—and numerous attempts to bottle them have been made. Thousands of references to the scents surrounding the experience of romance and sexual pleasure occur in literature. Some passages describe responses to the odors we label the "natural" scents associated with sex; many other references detail perfumes and the odor of food and wine or flowers that are part of sexual encounters—the odors we add to encourage or enhance our sexual experiences.

For Western culture, the Bible's Song of Solomon has provided rich sensory images that have been quoted extensively to illustrate the way sight, touch, sound, taste, and smell are linked to create sensuous comparisons between sex and the natural world.

> While the king sitteth at the table, my spikenard sendeth forth the smell thereof. . . . A bundle of myrrh is my well-beloved unto me; he shall lie all night betwixt my breasts. . . . His cheeks are as a bed of spices as sweet flowers; his lips like lilies, dropping sweet smelling myrrh. . . . My beloved is unto me as a cluster of camphire in the vineyards of Engedi. . . . I am the rose of Sharon, and the lily of the valleys. . . . As the apple tree among the trees of the wood, so is my beloved among the sons. . . . I sat down under his shadow with great delight, and his fruit was sweet to my taste. . . . The fig tree putteth forth her green figs, and the vines with the tender grape give a good smell. Arise my love, my fair one, and come away. . . . My beloved is mine, and I am his: he feedeth among the lilies. . . . How much better is thy love than wine! and the smell of thine ointments than all spices.

Apples, figs, wine, ointments, flowers—all appear in folklore throughout the world as sexual symbols, and some, such as figs, are associated specifically with masculine sexuality, but even more are linked with the feminine, which encompasses fertility and creativity as well as sexuality. Based on what we know about ancient and modern cultures, humans have always enhanced sexual life with abundant sensory stimulation.

The issue of adding odors to the sexual setting is often raised in a context that would sound very odd to ancient peoples or those from modern cultures other than our own. A woman in one of my lectures recently asked, "Isn't it 'cheating' to add scented candles or perfumed lingerie to lovemaking? Shouldn't love be enough to bring about the kind of mood we're trying to create?"

This question reflects some interesting attitudes that are prevalent in Western culture, but virtually unknown in many others. The first is the idea that love is a simple emotion that *should* automatically lead to sexual ecstasy. Logic tells us, however, that this isn't completely true. If it were, sexual-dysfunction clinics would be out of business, and the sexual encounters that occur between relative strangers would be unknown. Sexual drive and desire are much more complex than we like to think, and even the most happily married couples find that sexual desire waxes and wanes, often depending on other circumstances in their lives.

The question also reflects the idea that sexual life is separate and perhaps not as acceptable as "real" life—work, family, religion, and intellectual, artistic, and cultural pursuits. For as much sexual stimulation as we have in our society—and we are hyperstimulated by media images—we still isolate sexuality as an element of our nature that must be constantly watched over and monitored for "good" and "bad" expression. We also shy away from the idea that we need to put effort into creating sexual settings. If we love each other, we say, then we don't need all this fuss. But, looking across the many thousands of years of human history, this attitude is unique.

In ancient cultures and some societies existing today, spiritual and religious life were integral to every other part of life, not separated and compartmentalized, as we find in the West. In the *Kama Sutra,* an ancient Indian book on the "art of love," erotic life is described as a pathway to spiritual ecstasy, not a path away from it. Fertility rites of many ancient cultures were sacred rites, because the line between the spiritual and material was not only not clearly defined, but not an issue at all. Nature was sacred, and the senses—being part of nature—were sacred, too.

The ancient Chinese would find the question about odors and their role in sexuality odd indeed. Cheating? How could that be?

Pleasant smells are linked to affection, so of course scents are a part of the affectionate act of sex. The people of Pompeii would also puzzle over such thinking. When anthropologists excavated Pompeii, engraved on the walls of the sleeping quarters they discovered representations of pottery vessels holding fragrant oils. These surviving engravings offer a small but significant glimpse into the bedrooms of ancient people. The link between smell and sexuality was so direct and important that they depicted it in their art forms.

Ancient Greeks, Egyptians, Aztecs, and Romans made extensive use of odors in all their rituals, including those involving marriage, fertility, and the pleasures of sexuality. In many instances, the odors were linked to offerings to the gods of their cultures. (Almost all ancient people tried to please their particular gods by offering fragrances to "heaven," including those making offerings to the god of the Old Testament. Apparently, by whatever name we use for the source of creation, he or she has an excellent sense of smell.)

Because sexuality and spirituality were closely linked—and still are in many of the world's religions—using scents in courtship and mating rituals did not, for the most part, have negative connotations associated with it, as has happened in our more Puritanical society. Ancient Sumerians believed that scents were the ultimate seduction tool, and men were encouraged to use them to attract women. When I recently saw a young man hurrying down the street with a large bouquet of roses in his hand, the ancient Sumerians immediately came to mind. Some things don't change much. It's clear that the nose is a very sexy organ, and therefore, some people began to fret and worry about its power.

The Nose Must Be a Dangerous Organ

The father of psychoanalysis, Sigmund Freud, understood the link between sex and smell so well that he advised us to repress our sense of smell or we might walk around sexually excited all the time! (Perhaps he thought that people in the so-called primitive cultures were sexually aroused constantly because they used odors with such abandon.) No doubt a class bias reveals itself in what Freud was implying. The educated classes, those involved in what he considered intellectual work and "highbrow" art, were advised to sublimate their sexuality in order to rise above their "lower" nature. You see, Freud believed that our sex drive is dominant, but it must often compete with our ability to develop intellectually. According to Freud, it was evident that the lower classes were not suppressing their sex drive, or so he presumed, because they spent more of their time engaged in sexual activity and procreation than the more educated classes. (No one ever proved that was true, however.)

It is thinking such as this that permeated Western culture and combined with the power of religion to create concern over the senses and their power to lead us astray. If we didn't attempt to "overcome" our sensory drives, we could be led astray into "sin," as defined by our religious institutions, or into irrationality, as defined by the intellectual elite. Because sex and smell have long been linked, the misguided thinking that still influences us today led us to ignore our sense of smell. We would never encourage anyone to ignore the ability to hear or to see or even to taste delicious food, but we do behave as if smell is more potent in its influence. We're often cautioned about our sense of touch, of course, but we know it is essential to survival. Without it, we couldn't feel pain, for example, a signal that alerts us that we are hurt or injured. Thanks to Freud and others, we have been led to believe that our sense of smell is secondary and not essential for modern life. Therefore, repressing it has no significant consequences.

Havelock Ellis, another psychiatrist and sex researcher working in the early part of this century, delivered some mixed messages about the role of odors and sex.[3] On the one hand, he said that the role of odors in sexual life had been greatly underestimated. But the association among intellectuals with odors and "animalistic" and primitive behavior proved too powerful for Ellis, and he later modified his belief by saying: "If the sense of smell were abolished altogether, the life of mankind would continue as before, with little or no sensible modification, though the pleasure of life and especially of eating and drinking, would be to some extent diminished."

The influence of the psychoanalytic community was extensive and lasting. Twenty years ago, when I was in medical school, far less attention was devoted to studying smell and taste and the disorders and conditions affecting these senses than that given to sight or sound. Like others before them, my professors didn't think these senses were medically important, except in a few isolated cases, where odors indicated the presence of particular diseases.

We human beings are a resilient bunch, though, and in essence we have told the "authorities": no thanks, we'll keep on enjoying the smells around us. And the more we learn about our sense of smell, the more evidence we have that our survival as a species at least partially depends on it. The pleasures we derive from sexuality and the scents we use to enhance our experience are as important to our nature as breathing.

The Nose Makes a Comeback

As a scientist and physician, I can't rely on anthropologists and poets to prove my theories, but I can look to them to lead me into areas that deserve scientific attention and study. What has fascinated me again and again is that applying scientific methods to cultural myth or even poetic expression so often confirms what human beings have long known: that our ability to detect odors is not only pleasurable, but it may also be essential to our survival as a species. Nature never does anything frivolously, and the fact that we possess a sense of smell means that it is important.

We know that some odors are detectable in our surroundings, and hence, we can identify them and decide if we like them or if we should listen to the messages they may bring that signal danger—the smell of gas or fire, for example. We also know that some odors are just below the level of conscious detection, but they influence us in powerful ways. We term these odors "subliminal," and we'll examine how knowledge of this type of odor explains some often confusing human behavior.

Stress, Sex, and Being in the Mood for Love

Few people are able to "turn on" their sexual feelings on demand. If that were possible, sexual expression would never be a problem, and we know it is sometimes. In truth, the sexiest organ is the brain, and our mood and our feelings at any given time affect our ability to desire a romantic partner. Odors have a profound effect on moods, so it's beneficial to use them in our homes to help us create the atmosphere we would like to experience (more on this in chapter 8). But being in the mood to enjoy sex isn't a simple matter.

Starting in the late 1980s, physicians and psychotherapists began seeing the rise of a situation (some call it a condition) among married couples. (I suspect it applies to long-term partners, who for a variety of reasons may not be married, too.) Based on the common experiences these therapists and doctors heard about from their clients, the doctors labeled this phenomenon "diminished sexual desire." Apparently even young couples in their twenties and thirties voiced concern about their lack of sexual appetite. Even though young and healthy, they had lost interest in regular sex, something they had never imagined could happen to them.

Described as a symptom of modern life, diminished sexual desire seems to occur mainly among the affluent and successful—those our society calls "yuppies." However, as far as I know, no scientific verification that this group is unique exists. I think that these educated, younger people have simply been more willing to discuss their problems with their doctors and have sought marriage counseling or the advice of sex therapists more readily than other groups in the population.

In truth, a lack of interest in sexual activity can certainly be linked to the fast pace of our society. Most families have two working adults, and with that situation comes increased stress. In addition, the average worker in the United States is working more hours per week than ever before. When two adults, who may also be parents, are working a combined ninety to one hundred hours a week or even more, it is not difficult to see why their sex lives might suffer.

Recently, while at a health club, I talked with a man who described his ailing sex life this way:

We're up at six o'clock, with no time to linger in bed and even snuggle for very long. My wife heads to the shower while I get our two kids up. I start breakfast while she's helping the kids get ready for school. I check for urgent messages at the office, take my shower, and get dressed as quickly as possible, and then it's off to work. My wife goes to the health club for a quick exercise class before work and later one of us checks in with the afterschool sitter to make sure everything is okay at home. I never leave the office before six in the evening, and it takes me an hour to get home, so our evening is very short. We try to make dinner a pleasant family affair, but that's often hard, and if I'm late, the family has dinner without me. If I haven't done my workout after I leave the office, I try to run a few miles late in the evening, after the kids have gone to bed and the phone stops ringing. Some evenings we head to the school for some kind of activity, and my wife is a volunteer with one of the crisis centers in our town, so that takes her out of the house at least one night a week.

On the weekends, we try to spend extra time with the kids because the weekdays are so hectic, although sometimes one of us has to put in extra time at the office. After a Saturday night spent playing video games or going to a kids' movie, who has the time or desire for sex? I can't remember the last time we devoted attention to each other and made sex really special. We're always so tired and stressed out that sex is something we think of as a fantasy for our future, and we tell ourselves we'll have time for it when our kids are grown and our lives settle down. That seems crazy! It sounds like we're "saving" sex for our retirement years.

It may seem crazy, but it's also common. I hear couples lament that they have little time left in a day to set the stage for romantic evenings. Women complain that their husbands fall asleep on the couch in front of the television, and men complain that their wives are irritable and not in the mood.

Every now and then I'll browse magazine racks in bookstores and newsstands just to see what these publications are currently covering. Sure enough, I see articles with ominous titles that go something like this: "Is There Sex After Children?" or: "Where Has Our Sex Life Gone?" or even: "Is There Sex After Marriage?" This sounds bleak, and I don't know how pervasive this problem is or if it's just hype designed to sell magazines. But I do know that when life is hectic, playfulness and a sense of sexual fun often are the victims.

A study reported in the *New England Journal of Medicine* caught my attention because it explored sexual satisfaction among one hundred well-educated, middle-class couples, the majority of whom had been married for over five years—almost one quarter had been together for over twenty years.[4] This group was predominantly white, and of the one hundred couples, eighty-four had children, and 80 percent reported that their marriages were working well and were satisfying.

What stood out in the study, however, was that 40 percent of the men reported sexual dysfunctions, that is, problems with achieving and maintaining erections or difficulty with ejaculation. Sixty-three percent of the women reported dysfunctions, including difficulty with arousal or reaching orgasm. These types of sexual dysfunctions are considered physiological, and some people seek help from physicians or sex therapists for such difficulties.

Just as significant, however, were the reports about certain sexual issues the researchers termed "difficulties," which according to the couples surveyed contributed more to sexual dissatisfaction than the actual physiological dysfunctions. *And 50 percent of the men and 77 percent of the women reported these difficulties!* The types of problems mentioned in this category included complaints such as the partner choosing the wrong time to initiate sex, being unable to relax, generalized disinterest in sex, insufficient foreplay, feeling "turned off" to sex or even repulsed by it, and a lack of tenderness. Although the researchers aren't claiming that the one hundred couples in their study can be taken as completely representative of American marriages in general, the results do raise issues that probably are more common than we think.

From popular literature, such as magazine articles and self-help books, we know that women more often than men complain that their partners don't take enough time to arouse them and sometimes approach them for lovemaking when the mood and setting are "off." A woman I interviewed said that her husband read the paper or watched television while she cleaned up the kitchen after dinner, bathed their two small children, read them stories, and settled them down for the night. Then, just when she was ready for an hour of relaxation and winding down, he wanted sex and often approached her abruptly. And he wondered why she wasn't interested? "If he helped with these evening chores, I'd be more inclined to respond," she said; "but I resent that he sits back and relaxes while I'm doing the family's work."

I believe this husband should not only help with the evening chores, but he also should get his nose involved in creating a mood that is conducive to relaxation and sexual pleasure. What if the living room had the scent of vanilla or lavender candles? Wouldn't both partners end up happier if he prepared a hot bath with scented bath oil and invited his wife to join him in some unpressured time together? Perhaps in that setting, the husband would relax, too, and not be so hurried and abrupt. Women, more so than men, seem to understand that intimacy is more than joining two bodies together for a few minutes.

Imagine how different the results of any sexual-satisfaction survey would be if romantic partners took the time to relax together and added scents to their environment that would relieve their stress and induce a more calm, but alert mood. We know that certain odors cause us to be sleepy and others reduce anxiety and stress but keep us alert. Doesn't it make sense to use scents to our advantage—to enhance our lives by using our powerful and precious sense of smell?

You may have been using odors in a sexual setting all along; you may choose colognes and perfumes based solely on the fact that your partner likes them. Or you may use scented candles or potpourri in your bedroom simply because these odors make you feel good and help to lift your mood. It's likely not your imagination that your mood changes based on the odors around you. Odors are powerful. Think of it: simple molecules in the air can lull us to sleep or wake us up or put us in a romantic mood. Early evidence suggests that odors may also help treat some physical and psychological disorders. Most certainly, scents can enhance our sexual experience.

Using This Book

The information about odors that I describe in this book falls into three basic categories. One area includes the odors we can detect in the air and consciously use to enhance attraction and sexual pleasure. Oils, perfumes and colognes, and scented candles, linen, and lingerie are in this category of smells.

Another area includes the odors we can't detect on a conscious level. Pheromones (undetectable substances in the air—I discuss these in detail in chapters 2 and 3) and the unique individual odor we release from our bodies are included in this category, and these subliminal smells are very important. In fact, they may partially determine who we end up in bed with in the first place.

Still another category of odors encompasses the smells that alter mood and help create an atmosphere for romance and sex. We also know that some scents may not be conducive to sexual activity because they remind us, usually on an unconscious level, of something that prevents us from feeling romantic, or these odors may distract us and alter our mood.

All these categories of scents work together to create sexual experiences; anyone can use them, unless some medical reason exists not to. For example, if a person has a severe allergic reaction to some flowers, he or she won't fill the bedroom—or the living room—with them, nor should the person use perfumes that mimic floral odors. Some people with asthma will also need to exercise caution when experimenting with odors.

In discussions throughout this book I sometimes mention married couples to illustrate a point, and in other discussions I simply talk about partners or lovers. The information applies across the board, and it is of no concern to me whether sexual partners are married or not. Individuals with same-sex partners may enjoy the information and find it useful, too.

I don't want the information in this book to be misinterpreted as representing some kind of newfangled seduction tool that influences people against their will. No scent in the world is going to influence people to engage in sexual activity if they aren't completely willing; odors are not the same as drugs, and we can't make reluctant partners suddenly passionate and receptive by filling the room with odors. Even if we could do such a thing, it would be wrong, and we all know it. This book is written for consenting adults who act responsibly in their sexual lives and use safe-sex practices.

That said, let's move on. I believe that odors may be the *initiating* factor in sexual attraction, so consider your nose a partner in choosing a lover or spouse. That may be difficult to accept, but evidence suggests that attraction may start with the nose. One survey, conducted in Wales, reported that 71 percent of men and women rated smell as the *biggest* factor in sexual attraction. Smell was far more important than, for example, clothing and hairstyle—or lack of hair. The chemistry of love has baffled every philosopher, poet, and psychologist who ever studied the subject, and certainly our sense of smell can't explain all the mysteries of love and sex, but it can shed some light on the subject.

In the next chapter, we'll explore just how odors work on the brain and why they appear to be so important in our sexual lives. A basic understanding of the way your sense of smell works will help you put the information in this book to work, or I should say to play, in the bedroom—or any room for that matter!

2

The Irrational Nose

ROMEO AND JULIET are two of literature's most famous—and most analyzed—lovers. They, and so many other famous couples of myth, literature, and history, loved with abandon, and many would say unwisely. They met a tragic end, despite the attempts of their elders to do everything they could to keep them apart. Their elders warned Romeo and Juliet about the dangers of their liaison and tried to convince them to walk away from each other. After all, their families were bitter enemies, and no good could possibly come from their powerful attraction. Nevertheless, these young lovers became the most famous victims of dueling dysfunctional families—to put a modern spin on it.

Now, we don't know why these adolescent lovers were attracted to each other, but a number of theories can be postulated, based on modern psychology and science. Most psychologists would tell us that they were afflicted with a classic case of teen rebellion. If they'd only listened to their parents. . . . Well, short of that, if they'd given their relationship a bit more time, they would have come to their senses and realized that theirs was only a serious case of puppy love, and eventually they would have matured and married more suitable partners. (Suitable as defined by their families, of course.) There might be some truth to this version, but it isn't nearly as romantic as the notion of a great, uncontrollable love—a magnetism more powerful than any intellectual reasoning could overcome.

Those involved in studying the science of attraction would probably say that the two lovers had a strong visual attraction, one that convinced them that they couldn't live without each other. We do know that visual attraction is extremely important, at least initially. The less romantic among us would say that if their families hadn't been so impossibly stubborn, the pair would have realized that a visual attraction does not a relationship make. If they'd stayed together long enough they might have realized that they couldn't abide each other's tastes and personal habits. Or, had they married, they might have soon started fighting about money or housework, and maybe they would have had a rancorous divorce.

The more romantically inclined would say that Romeo and Juliet were simply the classic, star-crossed lovers, and while their story is tragic, fate led them to each other's arms, and now we should picture them living happily in eternity, without petty families to deal with. In other words, this is the ultimate romantic—and romantically foolish—fantasy. Die young, and love will never die.

Enter the Nose, Stage Left

I have a different theory about this mythic couple. I have a sense that their attraction originated in the nose. They were attracted to each other's scent, and the attraction was so strong, they couldn't stay away from each other. One can't see a stage or movie version of this story without wishing for another ending. We want them to be careful, to be rational, to find a safe way out of their dilemma, but our lovers continue their journey to doom as if they have no power to do otherwise. Yes, the sense of smell is very powerful indeed, and because it is closely linked with our emotional responses, it adds to the irrationality of love. Being young and impetuous, this pair was anything but rational.

Do you know couples that appear quite happy with each other even though people around them, including friends and family, can't understand why? Do you know couples that stretch even the extreme of the "opposites attract" truism? Perhaps you know a couple who bickers a lot, but they can't bear being away from each other for very long. They confuse everyone around them with what appear to be mixed messages about their feelings. What do they *see* in each other? As we gather more information about the sense of smell, we might ask, "What do they *smell* in each other?" When we understand the way our sense of smell actually works, this doesn't sound like a silly question.

It's in the Air We Breathe

The story of attraction starts with the simple act of inhaling. Unless we are in an environment that has been deliberately deodorized, we inhale odor molecules in the air every time we take a breath. Most of the time we don't pay much attention to our breathing, but if we detect a particularly unpleasant or even dangerous odor in our environment, we tend to deliberately take short, shallow breaths. On the other hand, if we smell dinner cooking and we're hungry, we usually inhale deeply to take in more of the aroma.

As we breathe, air currents develop in the nose, and although it's not important to go into great detail about this, one nostril is generally slightly more stuffed up than the other. This changes every eight hours or so, and we call it the olfactory cycle. Oddly enough, the air currents are stronger in the nostril that's the most stuffed up, and this nostril has better olfactory ability at that time. (If the nostrils are completely closed because you have a cold, your ability to detect odors is obviously decreased. Don't confuse the congestion associated with illness with the normal stuffiness that we usually aren't aware of.)

As air currents develop—I like to think of them as tiny tornadoes —in the nose, the odor molecules are able to reach the smell center located at the top of the nose, just behind the bone we call the bridge. The smell center contains epithelia, which are protective, mucous-coated membranes about the size of a dime. As the odor molecules move through the thin mucous membrane, the odor makes its way to a pin-size area where millions of receptor sites are located. These receptor sites are important because they are matched with odor molecules and allow us to distinguish between one odor and another.

When an odor molecule reaches a receptor site, the body's electrical system begins to play its part. While this is a complex system, just understand that at this point the odor molecule is changed so that it now becomes an odor signal that is intensified by a factor of about one thousand. Once the intensified signal is formed, the brain can respond to the odor.

It's important to realize that the pathway a molecule takes from the air to the brain is direct. Unlike the processing structures of our other senses, the olfactory nerve is projected directly from the brain, and the external stimuli—odor molecules—have direct access. The ear has the eardrum, which acts as a barrier outside the brain between the auditory processing system and sound; the eye has the cornea, which acts as the barrier between visual stimuli and the visual processing system. Our sense of smell is the only sense whose pathway to the brain is unencumbered, so to speak. This becomes an important factor in understanding why smell is the most "irrational" of the senses.

From an Odor to a Feeling

Our sense of smell is the only sense that is initially processed in the limbic lobe of the brain, which is the emotional center of the brain. All other sensory stimuli are processed first in the thalamus—the part of the brain that acts as a relay point between the cerebral cortex, the area of the brain that controls thinking and logical reasoning, and the sensory organs, which first register the sensations. For example, in contrast to smell, visual stimuli pass through the thalamus and the cerebral cortex. After being processed in the cortex, the limbic system is finally affected. This means that our thought processes and ability to reason get first crack, in a manner of speaking, at processing what we see. The same is true for our auditory apparatus and our sense of touch.

One way to think of this is in a sequence that seems artificial, but illustrates what we mean. If you see a picture of a house, your brain first tells you that it is a house, and then you decide if you like it or not. The processing system identifies the concept of "house" before you begin to make evaluations and judgments about it. But the nose doesn't wait for identification. When you detect an odor, your emotional response tells you if you like it or not before you even know what is causing it.

Obviously, this process goes on all the time, and we aren't conscious of it. For example, we do not experience a break in the sequence, and we don't say to ourselves, "This odor molecule is now in the smell center and will soon be intensified. Oh, there it is, my brain has it now." Sensory stimuli are always continuous and usually not consciously perceived.

You sometimes become conscious of your reactions to odors, however, because these emotional responses are often so swift. For example, you're walking down the street while thinking about your checkbook balance, but suddenly you pass a bakery, and all thoughts of finances disappear as you start a debate with yourself about whether to stop for a pastry.

If a particular odor has a strong emotional connection with a person in your past or present life, you could be overwhelmed by a flood of emotions. You might even experience waves of nostalgia. In that case, a particular odor has caused a poignant or bittersweet response, and if you're with someone else, you may find yourself telling the other person about the odor and what it means to you. Let's say you are in a restaurant and a woman walks by who is wearing a particular perfume. Immediately, you become a bit sad, and powerful, nostalgic images of your grandmother come unbidden to your mind. "That perfume is what my grandmother always wore," you tell your dinner companion. "I haven't smelled it for years, and I miss her so much." This feeling usually passes in a minute or two, but you may think about your grandmother later that day or find that she's on your mind for several days.

Any odor potentially has this power, not just a particular perfume. A patient, whom I'll call June, once told me that the smell of the horses that draw the tourist carriages through the streets of downtown Chicago reminded her of her life with her first husband. She had been walking from the train to our office, and this was the first time she had seen the carriages or inhaled the odor of the horses. Her eyes became teary when she told me about this because she hadn't thought about her first husband except in passing for quite some time. It was clear that we needed to talk about this before I began her medical examination, so I asked June to describe the experience and the memories in greater detail.

June told me that she and her first husband had grown up in a rural area, and their high school dates often included riding their horses together. The smell of the barn and the odor of the horses all came back to her in an instant, and immediately she was thinking about all the good times they'd had together. June had eventually married the young man, but two years later he was killed in the Vietnam War. Shortly after, she moved to Chicago to attend college, and within a few years, she had remarried and had two children. But, she told me, she never rode a horse again.

June volunteered that she often enjoyed the sight of horses and horseback riding in movies and such, but these visual images didn't affect her in the same profound way the odor had. In fact, the experience was upsetting because she felt so guilty about having the response. She thought that these images of her first husband were somehow disloyal to her current husband, with whom she'd shared a long marriage and a good relationship.

When it comes to the nose, it does us no good to say, "I shouldn't feel this." Our system of processing odors is designed to react first, identify later. Many people whose first love has not died become nostalgic when an odor reminds them of a special time in their lives. I've heard people comment about the smell of a gymnasium where they went to high school dances, and soon they say, "I wonder what ever happened to. . . ." They may even have a longing look in their eyes. These responses are part of life and we cannot do not much about them, nor should we want to. We can learn a great deal about ourselves when we allow the memories to surface and then examine their meaning for us.

By the time June left our office, I think I had convinced her that she was not being disloyal to her second husband so much as she was acknowledging a particular time in her life, one she associated with first love and carefree youth. That period in her life had ended abruptly and tragically, but she had moved on and created a good life. She wasn't any more stuck in the past than any of us is when an odor reminds us of people and places from years gone by. While not always pleasant, these odor-invoked nostalgic responses enrich our lives—if we let them.

As we'll see later, an understanding of the meaning of emotional reactions to particular odors can even influence our sexual responses to our partners or a particular setting or atmosphere.

Looking at the Nose and the Brain Another Way

Nowadays, some people better understand the workings of the brain if we use the terms "right brain" and "left brain." You may know that the left side of the brain controls intellectual processes and functions such as performing math problems and reading. The right side of the brain is dominant in activities we associate with creativity and artistic endeavors. Touch, sound, and sight are processed in the left side of the brain, the center of rational thinking. Smell is processed in the right side of the brain, the center that controls emotion and creativity.

All of us engage both sides of the brain simultaneously, and we should never conclude that one type of function is more important than the other. The so-called left-brained scientist is constantly using the right side of the brain because intuitive impulses and creative approaches to problems originate there. The right-brained artist uses the left side of the brain to organize what he or she has created. It's that simple—and of course, that complex, because the brain is an amazing control center every moment we're alive.

The next time you find yourself particularly emotional and don't understand why, consider that the odors in your surroundings may be acting on your right brain. In fact, our sexuality is certainly dominated by the right brain. We may make rational choices about our partner, but our feelings of affection and our sexual attraction are right-brain activities.

From a physiological viewpoint, our sense of smell is the only sense that substantially changes in tandem with reproductive development. For example, at puberty, odor (and taste) preferences change, as does our ability to detect certain sexually related odors.[1] Women's sense of smell is better at ovulation and is at its worst during menses. By contrast, changes in our ability to see and hear are not linked with puberty or with reproductive cycles.

Why Is Our Sense of Smell So Different?

Our sense of smell may *need* to be linked with our emotions because the detection-response mechanism is necessary for our survival. We haven't always lived in safe houses with plentiful food supplies. Over our evolutionary history, our sense of smell has been able to alert us to potential dangers we could neither see nor hear and most certainly not touch. The ability to detect the presence of an animal could have meant that dinner was close by or that we had to escape from the area because danger was lurking. Our ancient ancestors could probably smell herbs and edible plants hidden in other foliage far better than any of us could today. The act of yawning draws odor molecules to the back of the throat to the olfactory bulb. (The route for moving odor molecules to the olfactory center in the brain is called the "retronasal pathway.") Yawning before sleep may be the way our ancient ancestors detected the presence of an animal that could prey on them while they were sleeping. In the past, our sense of smell was essential just to stay alive.

Our sense of smell is relatively weak compared to that of other animals. A dog often seems like little more than a nose attached to some fur. Cockroaches can smell many thousands of times better than we can, which probably has something to do with the way they manage to survive despite our efforts to eradicate them. Our sense of smell may also be weaker because we no longer consider it so crucial to our well-being and, ultimately, our survival. As other senses became dominant, we used our sense of smell less, and it is no longer as "sharp" as it once was. Those whose other senses are impaired, however, tend to rely on the sense of smell. Helen Keller, who was able to live life fully despite being deaf and blind, often spoke eloquently about the power of odors. For her, smell, taste, and touch were her entire sensory universe. She was able to identify people by their odors in a conscious way; we often don't think about this, but when tested, many individuals, women more often than men, can identify their family members by their odors.[2]

In Western society, the importance of the sense of smell is given little attention, although I believe early human beings knew exactly how critical it is to survival. Because emotional responses to odors are so powerful, this was even more incentive to attempt to repress our sense of smell. Remember that our culture is dualistic and compartmentalizes many things. We almost always consider it necessary to label something as good or evil, and we tend to put many functions, even those rooted in the brain, into hierarchies. The intellect is elevated to a position above emotions, which is why so many people try to repress their feelings.

"Irrational" is not considered a positive word. My patient June, who I mentioned earlier, was embarrassed by her response to the horses. She had a difficult time accepting that she couldn't—and shouldn't—try to rationalize her response and explain it away. But that's what many of us have been trained to do.

In the sexual arena, we often feel guilty if we're attracted to a particular "forbidden" person. Guilt is certainly what Romeo and Juliet's families would have preferred, and perhaps the one thing that these two families could agree on is that a dose of guilt is better than an impossible love affair any day. Of course, it's necessary to have these controls on our behavior. Most of us don't jump into bed with every person to whom we feel attracted. But when we understand why we might experience these powerful attractions, we can allow ourselves some leeway and stop feeling so guilty. The reason for these "out-of-the-blue" emotional responses to a particular person may start in the nose, an organ we can't do much to control.

You might be thinking that you can't link your attraction to your partner to any particular odor. You would say that you fell in love with the person because he or she was your visual type. You like the way your partner looks. Then, you began to respond positively to the individual's personality. How could odors have anything to do with it?

In the chemistry of attraction, the most important element may be the odors you can't consciously detect, and that's what I think happened to our star-crossed lovers. We could say that their hormones got the best of them, but we might want to look at those invisible pheromones first.

When Love Is in the Air

Pheromones are substances we can't see and don't consciously detect, yet strong evidence suggests that they are in the air around us all the time. The reason we aren't affected or influenced by all the pheromones in the air is that they aren't meant for us. By that I mean that pheromones are species-specific. The animal world's mating habits are virtually ruled by these substances.

From studies conducted on other animals and insects, we know that pheromones are odorant substances that one sex of a species releases. The other sex of the same species is influenced by these pheromones for the purpose of mating. A female moth releases a pheromone that male moths can detect a kilometer away. The female moth's pheromones don't attract ants or butterflies or beetles. They attract only the appropriate mate for female moths. We can take a walk on a summer evening and not be distracted by the insect pheromones that fill the air. It isn't a question of ignoring them; they have no power to influence us.

Dogs and cats are acutely aware that a female of their species is in heat—you are aware, too, if you happen to live in the neighborhood. All that barking and screeching going on is the natural response to the pheromones being released by the female dogs and cats. It amuses me to see pet owners' annoyed response to the animals that attempt to come around to mate with their house pets. As with farm animals, we have dictated the breeding habits of pets to the point that people are sometimes upset when their male dog raises a ruckus when a female dog passes by. To some people this is "unseemly" behavior, and I even heard one man tell his dog to mind its manners and behave. The dog *is* behaving, exactly the way nature intended him to.

Based on what we currently understand, pheromones are part of nature's design to ensure that every species procreates. And as pheromonal science has progressed, we are coming closer to establishing that they exist in humans too. Until we can definitively isolate the actual substances, however, we can only use the strong body of evidence we have to infer that they are in the air.

The Strong Case for Human Pheromones

When we use pheromones in human experiments, we call them "potential" pheromones because we can't definitively say that the substances we use are the real thing. However, when these substances are used, human behavior can be altered, which leads us to believe they are pheromones.

We believe that pheromones are by-products of the hormones produced for reproduction. Both males and females produce androgen, estrogen, and progesterone. Women produce more estrogen and progesterone, and men produce more androgen. Therefore, the pheromones each sex produces are different.

The evidence suggests that pheromones are released by special glands located in the underarm area and around the genital organs, as well as other places on the body. These are the apocrine glands, and they produce high-density steroids. (At one time, these glands were thought to have no function in humans.) They are not to be confused with the eccrine glands, commonly called the sweat glands, which are part of the body's cooling system and may also activate when we're nervous or anxious. The eccrine glands are stimulated by physical exertion, heat, and a generalized reaction most people have to stress. The apocrine glands, however, are more sensitive to mental stimuli, particularly sexual cues. (They may also activate in situations that are perceived as threatening and fearful.) Prior to puberty, these glands are not active, but they later become important in the sexual development in both males and females. Their proximity to the eccrine glands may be nature's way of distributing the apocrine gland secretions through sweat.

Humans have more apocrine glands than any other animal, and they are larger in size. Male apocrine glands are larger than those found in females, but females have 75 percent more apocrine glands than males, which is another indication of their function as part of sexual lure. Their wide distribution in the axillary—underarm—region may be linked to the fact that we stand upright, which favors the upper body as the center of odors associated with sexual attraction. Among primates, humans are the only species that mate in a face-to-face position, which allows the nose ready access to the underarm area.

In Elizabethan times, there was a custom that usually sounds strange to many people in our culture today. A woman peeled an apple and carried it under her arm, thereby allowing her natural odor to permeate the flesh of the fruit. She then gave the apple to her lover, which enabled him to be close to her scent—and of course, he could ingest it. In light of pheromonal science, perhaps it was the odor of pheromones that was the lure.

In our culture, most women shave off their armpit hair, but the physiological function of the hair may be to trap apocrine gland secretions in the armpit cavity. Then during sexual stimulation, the alluring odors are released. The scent of pheromones may also be released when the arms are raised. Those knowledgeable about body language tell us that raising the arms, as in folding the hands behind the head, is a gesture of openness to another person. In a romantic context, it may also be a gesture that signals sexual receptiveness.

When a woman raises her arms to embrace her lover, it's possible she's releasing an odor that is meant to attract him—and often does. Dancing also involves raising the arms. Beginning in the 1960s and continuing today, much of the dancing done by young people (and now the middle-aged, too) doesn't involve touching, but if you watch carefully, much dancing involves raising the arms, sometimes high over the head. During intercourse, many women raise their arms behind their head, a gesture and signal of passion, but also, perhaps, a release of pheromones. In early courtship, putting an arm around a woman can be a sexual signal, a relatively safe first step in expressing romantic feelings and a "test" to determine responsiveness. Unconsciously, it may also be a way to release odor molecules that act to attract the woman. If the woman likes the odor—and the man—then sexual activity may progress.

On the other hand, women often note that what was meant to be an affectionate hug is often interpreted by a lover as a sexual overture. Unconsciously, a man may be influenced by a pheromonal odor, and indeed, he begins thinking about sex. Perhaps he didn't have sex on his mind either, until those alluring pheromones were released. Perhaps miscommunication among lovers is at least somewhat biologically based.

Male and female body chemistry is quite different because the reproductive hormones, and presumably the pheromones, serve slightly different functions. For example, cross-gender research about olfactory acuity, or to put it more simply, the degree to which we are able to detect and identify odors, reports that women have a better sense of smell than men.[3] (If a woman tells you the house is burning down or the garbage is beginning to smell like last night's fish, I'd pay attention. She is usually right.)

A woman's sense of smell is particularly heightened during ovulation, which is probably of evolutionary importance. If nature is encouraging procreation, it makes sense that women's ability to detect pheromones released by the males around her is better than it is during menstruation. Even though she may not be conscious of the male pheromones, she is often more receptive to mating during the ovulatory phase of her cycle. Many women report increased sex drive during the ovulatory period in their cycles, and this is probably no accident. It's been reported that some women not only have more frequent intercourse at mid-cycle, but experience more orgasms during that time as well. (Some women also report increased sexual drive during menstruation, which may have psychological significance. During that time, a woman often believes she is not as likely to conceive and, therefore, feels safer having sex. While a woman is statistically less likely to conceive at that time, it remains possible, however; so the days of menstruation should never be considered a time when contraception is not needed.)

Nature is constantly urging us to procreate—that's nature's job.

However, we don't live entirely by nature's call anymore, so we have devised birth control methods to allow us to enjoy our sexuality while minimizing the risk of pregnancy. Obviously, this isn't a perfect system, or every pregnancy would be planned. But most people in our society today use some means to prevent conception without abstaining from sex.

Unlike many other mammals, we are also not dependent on fertility cycles to trigger sexual desire and activity. Some higher primates are also more like humans in this way, so we are not unique in this regard. The term "estrus" describes the situation in the animal world in which the female is receptive to mating, and of course, this span of time coincides with fertility. In our evolutionary past, estrus has been suppressed, allowing us to establish societies that are markedly different from other mammals.

Does the Family Depend on Pheromones?

It is likely that the reason we can even attempt to have a family system, especially a monogamous one, is that pheromones are in the air continually, and sexual relations are possible at any time. Once we're attracted to another person's odor, even an odor below the level of conscious detection, a strong sexual and emotional bond is possible. In other words, our evolutionary sexual development probably has much to do with the fact that we can sustain relatively stable family units. They aren't perfect, of course, but think how differently human society would have developed if mating was possible for only a short time each month—or only a few times a year. Some sociobiologists have concluded that this ability to form a stable bond was necessary because of the length of time it takes to raise our offspring to maturity. Because of this long-term dependency, it was important that males stick around and be part of the family group. (Again, this is obviously not a perfect system and, in fact, must be reinforced with social controls, or it breaks down to an extent.)

While we aren't ruled by pheromones, neither should we discount them when we attempt to explain human behavior. Those mismatched couples we often puzzle over may have really liked each other's pheromones, so much so that all the logical, rational reasons they shouldn't be together were overcome.

The question of homosexuality comes up, too. We do not understand why some individuals are attracted to the same sex or are bisexual if, theoretically, pheromones create sexual distance between those of the same sex. There is much about pheromones (and sexuality, of course) that we don't understand but, as yet, pheromonal science does not shed light on the reasons some people prefer same-sex relationships.

Pheromones Aside, My Partner Just Smells Good

While we know that pheromones are not detectable at the conscious level, we also are aware that each person has an individual scent. Perhaps the pheromones play a role in creating this scent; we aren't certain about this yet. But we do know that even in the absence of artificial odors—perfumes, powders, aftershave, and the like—individual scents are powerful in the dance of attraction. Perhaps what really happened to Romeo and Juliet is that their youthful pheromones provided the first tug on their heartstrings, and their individual odors were so powerful that they let their good sense be driven underground. If either of them had smelled just slightly different, there might not have been a play to write.

Since pheromones play a large role in sexuality and attraction, it helps us understand the chemistry of love if we learn more about them. In the next chapter, we'll look at the way in which these invisible substances influence you when you're not paying attention.

3

The Pheromones among Us

IN THE EARLY 1970s, Martha McClintock, an undergraduate psychology student, noticed that the menstrual cycles of female students in her dorm began to synchronize not long after they began living together. The length of their cycles began to shift so that within a few months they were in the same phase of their cycles at the same time. As a research project, McClintock decided to document this phenomenon and found 135 women (ages seventeen to twenty-two) willing to be research volunteers.[1] Some of the women were close friends, and some were roommates. For five months, McClintock charted their cycles, plus those of some randomly chosen women who did not know each other. As the months passed, the menstrual cycles among women who knew each other or were close friends moved increasingly closer to synchrony. The randomly chosen women showed no change in their cycles.

Women had noticed this phenomenon before among close friends and office mates, but before McClintock charted the data, the phenomenon had remained in the realm of anecdotal information and had failed to raise significant interest among researchers.

Many questions remain. Why does menstrual synchrony occur? Is there any significance to the phenomenon? Does it tell us anything about pheromones? McClintock's research clarified that menstrual synchrony wasn't a random event. It appeared that the phenomenon was the result of the associations between the women, specifically the amount of time they spent in physical proximity. In other words, the closeness of their friendship mattered less than the amount of time they were in each other's presence.

McClintock also documented that women who rarely dated had longer cycles; the women who dated regularly had shorter cycles. A study performed at the University of Pennsylvania applied an extract from the armpit of a male to the upper lip of women; the application of this substance showed a stabilizing effect on women's cycles.[2] In other words, the menstrual cycles of the treated women tended to be more regular than those of the untreated women, suggesting again that the presence of males influenced cycles.

It was clear that the women in McClintock's study were influencing an important and measurable biological function. In addition, their relationships with men apparently influenced their cycles, an observation confirmed by the later research. This is really quite remarkable because women are unable to consciously regulate menstruation. Women can't decide when they're going to menstruate or control how long their cycles will be. Later, other research corroborated McClintock's findings, which was important because replication is necessary before we can draw definitive conclusions.[3] But, if a substance secreted by males in the underarm region can alter women's cycles, then the chemical itself represents a *potential* noninvasive treatment for menstrual irregularity. Obviously, more research is needed to isolate the substance and determine how best to use it, but research such as this may have great implications for human health.

About the same time that McClintock's research was published, a scientist in England, a man who lived in isolation much of the time, began an experiment that sounds odd in the telling but that may be of significance to researchers studying potential pheromonal effects.[4] After shaving every day, this scientist measured the dry weight of his beard growth, the hairs he shaved off that were left in his electric razor. Periodically, he left his isolated home to visit with his lover and be in contact with other women. When he was with his partner and around other women, his beard grew more rapidly. This indicated that being around the opposite sex had an effect on beard growth, which is linked to production of the male hormone, androgen.

The Theory of the Dominant Female

Once menstrual synchrony was documented, unanswered questions still remained about the reasons for this physiological phenomenon. In one research project, conducted at Sonoma State Hospital, a woman— we'll call her Sue—noticed that other women seemed to synchronize with her cycle.[5] To see if this was true, Sue was asked to wear sterile cotton pads under her arms to collect her perspiration. Then, three times a week for four months, researchers applied a solution containing her perspiration to the upper lip of eight women, none of whom was acquainted with Sue. A control group of eight other women had a "placebo," an inactive substance, in this case a simple alcohol solution, applied to their upper lips the same number of times.

The results were not ambiguous. The cycles among the women in the control group didn't change at all, but the cycles among the women who received Sue's perspiration began to move in line with her cycle. The implication is that there is something about Sue's odor that has the ability to *unconsciously* influence hormonal production in other women. This theory also applies to women in dormitories and offices. One dominant woman may influence others in her group, and the chemical agent by which she does this is hypothetically a potential pheromone. No one knows the way in which "dominant" is defined in this context, and women have always been considered less territorial than men. But these research findings raise questions. For example, if you randomly chose a different woman, and used her perspiration, would the results have been the same? Or is there something about Sue's pheromones that tend to hormonally dominate other women? In a dorm setting, are the women's cycles "trained" by a woman who, for unknown reasons, is a natural "hormonal" leader? And do the women who dominate in this way have a personality profile different from those whose cycles synchronize with hers? At this point, we don't know the answers to these questions. It is possible that dominance, in the traditional sense of the word, has nothing to do with the phenomenon, but is instead related to body chemistry alone.

Detecting the Chemicals in the Air

If we don't know for certain that human pheromones exist, then how do researchers know what substances to investigate? As in many areas of science, other animals offer clues. For example, androsterone and androstenol are chemical compounds, specifically, high-density steroids, found in the saliva of hogs of both sexes. Hog farmers use a product derived from androsterone to induce the females into lordosis, a posture needed for mating with boars. These same chemicals are present in humans and are secreted from apocrine glands located under the arm and in the groin area and in other parts of the body; traces of androsterone are also found in human saliva, suggesting that kissing is just what we thought it was, a way to form and express a sexual bond.

A group of researchers at the University of Texas in Galveston were investigating the ability of androstenol to influence human behavior and discovered that, indeed, it did.[6] When they placed androstenol in some of the stalls in a men's rest room, men avoided those stalls. This suggests that the presence of a male pheromone was unconsciously processed as a territorial marker.

In another study, androsterone was placed beneath a chair in a dentist's waiting room.[7] The researchers found that women tended to choose that chair to sit in as they waited, but men avoided it.

Both of these studies involved a type of forced choice, and the data are objective. When you enter a waiting room with many empty chairs or a rest room with empty stalls, you are forced to choose one, although you may not consciously use a rational decision-making process. Other studies have attempted to elicit subjective evaluations based on the presence of pheromones.

This Woman Is Really Attractive!

Can the presence of androstenol influence evaluation of women—women who aren't even in the room? In a study conducted in the late 1970s, researchers placed a pheromone on surgical masks to discover how men and women responded to photographs of a variety of women.[8] Two groups of men and women were tested. The first group wore plain masks, and the second group wore masks that were treated with androstenol. The latter group tended to rate the women in a more positive way than the group whose masks were not treated. Those influenced by the subliminal odor in the mask thought the women were more attractive and generally more sexually appealing. Remember, the only factor that was different was the presence of the potential pheromone, and both men and women were influenced by the substance. So, though it may be contrary to conventional wisdom, beauty may be in the *nose* of the beholder.

Both androsterone and androstenol are believed to have pheromonal effects.[9] About 30 percent of men and 70 percent of women can detect the odor of androstenol; about 92 percent of women can detect androsterone, but 44 percent of men are unable to detect its presence.[10] Both chemicals have a musky smell, and virtually all perfumes use a musklike substance as their base. Among those who can detect androsterone, some find it pleasant and "musky" and similar to the odor of sandalwood, while others find it unpleasant and say that it has a urinelike odor. More adult women than men found androsterone unpleasant. Responses to androstenol are similar, although among people who can detect it, some describe it as having a "flowery" smell in addition to being musky.

The odor of musk is interesting in itself, since women are about one thousand times more sensitive to the odor than are men. At the time of ovulation, women are one hundred thousand times more sensitive to musk than men! This sensitivity is lost when a woman's ovaries are removed but may be restored if the female hormone, estrogen, is taken. This suggests, of course, that reproductive hormones are intertwined with olfactory mechanisms.

While neither of these steroids, androsterone or androstenol, is a proven pheromone, some researchers consider androstenol the more powerful of the two substances. Both are extensively used in investigations of potential human pheromonal response.

In the plant kingdom, truffles are prized as a culinary delicacy in some circles, and they contain both androstenol and androsterone. Pigs are the best truffle finders, and it is probably the presence of these two steroids in the plant that leads their noses to the right place. Caviar, celery, and parsnips also contain the two chemicals, and caviar's reputation as an aphrodisiac is legendary. (As far as we know, this truly is a legend; no scientific proof exists showing that caviar is an aphrodisiac.) That many of the same chemical substances are produced in both the animal and plant kingdoms is not unusual. For example, soybeans contain estrogen, and a particular species of yam contains progesterone.

Various scents and tastes are used as sexual signals in animals, and they function to attract females and also to keep other males away. The sensitivity the females have to the odors depends on their reproductive readiness. If a female is not in estrus (better known as "heat"), then the odors the males are exuding won't attract her. Among some mammals, male odors may also induce estrus, which is powerful evidence that odoriferous chemical compounds have the ability to alter reproductive cycles.

Some researchers have speculated that a woman's breath changes during her menstrual cycle and acts as a pheromonal signal for males. Since potential pheromonal substances occur in saliva, this reinforces the idea that procreation is *unconsciously* encouraged. Changes in pheromonal substances that are the result of changes in reproductive hormones serve as a signal. The pheromonal activity may influence the woman, the man, or both.

Exalotide, a substance that is chemically similar to male sexual musk, is detected to a much greater degree by women than by men, and the sensitivity to this substance peaks at ovulation. A woman who has had both ovaries removed often loses her ability to detect the odor, but as with the odor of musk, the ability may return if estrogen is prescribed. This suggests that the ability to detect a potential pheromonal signal may depend on and fluctuate with the production of other hormones.

Aliphatic acids are among the components of vaginal secretions, and researchers have attempted to determine if this substance could influence sexual behavior among married couples.[11] The study volunteers were divided into four groups. One group of women in the study was given a substance containing this odor and instructed to rub it into their chests on alternate nights. The other women in the study were treated with one of three substances—that of water, of a perfume in an alcohol solution, or of alcohol alone. Sexual activity among all the couples was then recorded and analyzed by the investigators. They found that the married couples fell into two groups. Prior to the odor treatment, one group tended to have increased intercourse around the time of ovulation, and in that group, the presence of the vaginal odor resulted in even greater sexual activity. The second group tended not to be cyclic in their sexual activity in the first place, and the odor had no effect.

It's possible that among the first group, either the men or the women—or both—were responsive to pheromones, which accounted for the increase in intercourse during ovulation. If it was the men, this means that they directly responded to the female pheromones; if the women responded, then they developed a behavioral response that induced sexual arousal. And it's possible that both were aroused, and this increased sexual activity. This doesn't mean that the group was more fertile; it simply means that the desire for sex was stronger at that time. Within the second group, where no change occurred, it's possible that the men or the women are less responsive to pheromones. Perhaps only one partner responded to the pheromonal signal, but the other didn't, so sexual activity didn't take place. These remain unanswered questions at this time, but research appears to point to the existence of human pheromones.

But If Pheromones Don't Exist, What Causes Our Responses?

Alternate theories suggest that what we call potential pheromones are simply the odors associated with sex, and our response to them is conditioned. In other words, we detect certain odors in the air, link them with sex, and respond to them because we've been conditioned to do so. This would apply to subliminal smells as well as to those we can consciously detect. Our responses may involve an element of conditioning, but that doesn't fully explain, for example, the strong links between the changes in olfactory acuity that occur during the menstrual cycle.

Another theory maintains that smells, particularly male odors, are essentially unpleasant to females, and males must overcome their unpleasant odors before successful mating can take place. A study of mating patterns among rats found that female rats are able to distinguish the odors of males and reject them based on their odor.[12]

However, if the olfactory apparatus of the female is damaged, and she loses her ability to smell, she'll mate with any male. This supports the idea that male odors are important in helping females choose the best mates.

Another theory suggests that our own odors stimulate our own sexual excitement. Havelock Ellis said that it was possible that male sexual odors actually aroused the man, and the woman's response was stimulated by his sexually excited state. In other words, Ellis was saying that the male odor isn't particularly appealing to women, but increased arousal in men. Ellis expressed his theory early in the century, however, which was long before we had any knowledge of the way in which male odors can influence female menstruation, for example. However, the idea that odors are self-arousing, probably in both sexes, doesn't preclude the arousing nature of a partner's odor.

It is also true that odors alone do not necessarily lead to sexual behavior; other sensory cues are needed to complete arousal. For example, male hamsters will mount other males who have been treated with vaginal secretions. They are attracted to an odor that is processed as "female." But additional cues are present, perhaps both visual and olfactory, that are processed as "hamster." You see, the male won't mount a clay model of a hamster that has been treated with vaginal secretions. He knows that, regardless of the odor, this is not his species. This suggests that other necessary sensory "ingredients" are needed for a sexual responses.

Science often produces data for which we have no clear explanation. For example, androstenol production in men increases fivefold in late December, which coincides with the winter solstice. Babies conceived at that time are born in September, which is when food is the most plentiful in our hemisphere. Does this mean that the androstenol production increases to attract females at the very time that conception would be advantageous in terms of the food supply? We don't know. Late December is also the darkest time of the year, and we know that production of certain hormones shifts with the cycles of light and darkness, so perhaps the increase in androstenol is linked with circadian rhythms. I mention this because many questions about hormonal functioning exist for which answers are not apparent at this time. So far, we can only look at the pieces of the puzzle we have and draw tentative conclusions about their ultimate meaning.

As you'll see in chapter 9, the chemistry of attraction in the animal kingdom has long been of interest to the men and women who create perfumes and other scented products whose purpose is to enhance the dance of attraction. The producers of cosmetic scents and the scientists who study reproductive chemistry can't say for certain that they've isolated human pheromones, but the evidence points to the existence of important chemicals that influence our behavior. Nothing in nature exists without purpose, and it is unlikely that compounds with a primary role in reproduction in other primates are mere artifacts in humans.

Are We a Bunch of Zombies Being Led Around by the Nose?

Pheromonal research may help us understand human sexuality in a much deeper way. Remember that romance, love, and sex aren't dominated by the rational left brain; rather, the right brain, which dominates emotions and basic drives, is the major influence in this area of life. Thus, all societies create external controls and "rules" that are meant to govern our sexual behavior and guide us to appropriate mates. From time to time society changes its definition of what "appropriate" means, none of which matters much to the nose.

One reason the D. H. Lawrence novel *Lady Chatterley's Lover* created such a scandal was that Constance Chatterley's secret lover, and for all practical purposes, her first lover, was the gamekeeper on her impotent husband's opulent estate! If she'd been sexually awakened by a man of her own class, even one that wasn't her husband, the book would have lost some of its power to shock, sexually graphic though it was by the standards of the time. Constance broke the rules with an inappropriate man, and Lawrence explored the implications of her actions and the deeper meaning of responses to forbidden love. When you consider the power of pheromones, it may help you understand why it sometimes seems that you are helpless to resist a strong attraction. Some of the mysterious tapestry of emotions we call love begins to unravel when we learn more about our biochemistry. When you understand the "chemistry," then you won't be so helpless.

I think this is a positive development, because understanding helps us live more comfortably with the strange and powerful responses we may have to certain individuals. It sounds contradictory, but if we understand that we may, in fact, be led around by the nose, we can stop being led around by the nose—if we choose to. We can't alter what we don't understand, and we are better able to look for solutions to our problems if we comprehend the unconscious influences on our behavior. The visual components of attraction are so widely accepted that we don't even have a phrase in our language that implies that we're being "led around by our eyeballs."

Thirty years ago, to suggest that women's menstrual cycles could be influenced by a subliminal odor would have seemed preposterous. And would supposedly sophisticated males be influenced to avoid certain stalls in a rest room because of an odor they couldn't even detect? It seems impossible, yet it happened. Men who appear overly territorial are often viewed as "unenlightened," yet a territorial instinct appears to exist. Pheromonal science may go a long way to explain why this seemingly primitive behavior influences men, even when they don't know it.

A tendency exists to fear research that confirms our animal nature, but it is also humbling and gives us a glimpse into areas about which we still have limited understanding. On a lighter note, the biological similarities between humans and others in the animal kingdom is amusing and gives us a chance to laugh at ourselves. Who would have imagined that the nose was *this* important?

So, Are We Just a Bunch of Animals?

Most of the time your responses to olfactory stimuli are tempered by other sensory stimuli and the rational thought processes that separate you from other primates. If one chimpanzee runs off to mate with a new mate, that's not noteworthy. When Constance Chatterley risks her fortune and social status to mate with the gamekeeper, that's something to talk about. Society expects you to use your ability to make rational choices to override impulsive decisions based on initial attraction.

Pheromones exist throughout the animal kingdom, so evolutionary evidence shows that human pheromones also exist. Even if we don't call these smells pheromonal in nature, current research is confirming that subliminal odors have the power to alter hormone production and influence sexual attraction, drive, and even behavior.

The Changing Pheromones

Just as hormone production changes, we believe pheromones fluctuate as well. For example, during a woman's menstrual cycle, her body odor is stronger when estrogen dominates and is weaker when progesterone peaks. By body odor, I don't mean the odors we label "bad," or offensive, in our culture. I'm referring to the natural odor that is unique to each individual—our odor signature—that may be determined by the chemistry of lipids (fats) present on the skin.

Male body odor tends to be stronger because androgen dominates and interacts with bacteria on the skin. Males have larger apocrine glands and more body hair, which facilitates the diffusion of odors. It is likely that an increase in a woman's sex drive during ovulation is in part due to her increased ability to detect these male odors. So, if a woman is attracted to a particular man's odor, then her attraction intensifies during the very time she is biochemically most receptive to sex and when chances for procreation are optimal.[13] Nature's way may seem inconvenient in the modern age, but if survival of the species is nature's goal, and it most certainly is, then this system is exquisitely designed.

In both sexes, the body odor may actually change in response to stress, which probably affects sexual desire and response. In our language, we have expressions such as "the smell of fear" that acknowledge that something in the atmosphere—the air—changes when the fight or flight response is triggered. It isn't surprising that extreme stress may affect body chemistry and alter body odor.[14] Unconsciously, this may be a signal that we're not receptive to sex.

Can We Suppress—Or Attempt to Hide—Pheromones?

Based on knowledge of animal studies, we know that body hair is much like a net or web that traps pheromones. In humans, the glands that produce the substances we believe to be pheromones are produced by glands located in areas of the body, except for the head, that have the most hair. Hair has long been associated with sexual attraction, so it doesn't appear accidental that many cultures have customs involving women's hair and clothing.

In some cultures, women must cover their entire bodies, including their heads, any time they are seen in public; in other societies, women must cover their heads when they enter a church, temple, or mosque. Paul of Tarsus (St. Paul) is quite clear in his writings about the necessity for women to cover their heads in church or when they are praying in public. To do otherwise was to disgrace their husbands. At one time, Christian nuns were required to be covered from head to toe, without any hair showing, and women in some Buddhist religious orders also traditionally cover their hair.

It's possible that these cultural practices are unconscious attempts to suppress the pheromonal effect. Suppressing the chemicals of sexual attraction is a way to ensure that women aren't attracting inappropriate sexual advances. In addition, many cultures have feared female sexual power and have condemned women for their ability to "tempt" men or, at the very least, distract them from their male responsibilities; covering the head and body is symbolic of removing women's ability to lure men into "sin."

In cultures that more fully accept the senses and link them to nature and spirituality, men's and women's bodies were adorned, but not necessarily covered—beyond what was needed to protect against the elements. We have assumed that the covering of the body and hair was designed to minimize *visual* sexual cues, but in light of pheromonal research, we may have to revise our thinking. If the initiating sense of attraction is actually smell, then the unconscious motivation for regulating women's attire may well be to suppress pheromones.

While we can't wash away our signature smell, pheromones can be washed away. Before daily bathing was a common practice in Western culture, women's clothing was typically designed to cover much of the body; it was also unacceptable for men to appear in public with their chest or legs exposed. Today, however, we typically shower and wash our hair once or even twice a day, and pheromones no longer have the opportunity to build up and be caught in the web of the hair on the head and body. Throughout this century, women have gradually begun to wear less restrictive clothing, which also covers less of the body. Clothing requirements for men have also changed, and a bare-chested man at the beach or the ballpark—or even on the street—is now common. Is this coincidental? It could be, but I doubt it.

Remember, too, that in most of our society, women shave their armpit hair, and many women wouldn't think of wearing sleeveless clothing with even a hint of underarm hair growth showing. The underarm hair is much like an odor diffuser for apocrine gland secretions, which are linked with sexual signals. Perhaps removing the underarm hair is the Western equivalent to requiring women to cover their bodies in many Middle Eastern countries.

It's likely that the quantity of accumulated pheromonal messengers has considerably lessened as a result of our hygiene, so we no longer perceive a need to suppress their power. The body's production of pheromones probably hasn't changed, but removing body hair, combined with our daily shower, is doing the job once accomplished by more modest clothing. Conversely, now that we do "dispose" of pheromones, we may need more flesh exposed in order to have sufficient pheromonal distribution. But this still doesn't explain why we are attracted to certain individuals and not to others.

Is There a Biological Reason Opposites Attract?

An individual's odor signature is determined by one's genetic profile, or genome. Over time, successful procreation of our species depends on minimizing the risk that genetic weaknesses, or "defects," are passed from generation to generation. You don't mate with close relatives because their genome would be too similar to your own. More birth defects and other anomalies occur when the gene pool is small because of increased risk of passing on the same genetic errors. Therefore, a selective advantage comes with being able to identify differences and similarities. The odor signature acts as a signal that the person to whom you're attracted has a different genome, and in turn, successful procreation is more likely. Your genome is unchangeable; your diet, personal hygiene, or use of an abundance of perfumes or other scents doesn't change your genome or your basic odor signature.

When you say that you don't have "chemistry" with another person, no matter how much your family and friends want you to be attracted to a particular potential mate, this could mean that the genome of the other person is too similar to yours, so the other person's odor doesn't attract you. And when you wonder why the oddball couple next door is so compatible, even though they appear to have nothing in common, their different genomes may be the reason. For evolutionary reasons, it makes sense that opposites attract.

The idea that you have a signature odor that strongly influences your choice of a mate doesn't negate all the other reasons couples are attracted to each other and form lasting bonds. People who seem to have all the acceptable reasons for choosing each other, such as the same ethnicity, religion, socioeconomic status, and so forth, may have very diverse genomes. In this case, the external factors seem obviously compatible, and biologically, their genomes increase the chances for successful procreation. Sometimes, even though the external factors—the "demographics"—seem mismatched, the genomes could be ideal.

Psychological reasons also explain why opposites attract, and these theories have dominated our assumptions about the way in which we choose mates. For example, introverts frequently marry extroverts, and the theory is that they seek in the other person what they sense—usually unconsciously—they lack. Psychologically, the extrovert longs to be self-contained and less in need of other people for stimulation. The introvert wants to overcome social reserve and be more comfortable in the company of others.

When two people talk about "complementing" each other, this is related to the unconscious idea of coming together to produce a strong, complete unit that has everything it needs to be whole. I'm not saying that this is either good or bad, but compatibility doesn't require sameness, and in fact, many people are much more "opposite" than they initially believe. The first few years of marriage may reveal all the personality differences that aren't initially apparent. Ultimately, demographic similarities may mean little when it comes to lasting attraction.

But What about Love in Cyberspace?

Before we had Internet romances, we had mail-order brides, and men and women who fell in love through letters. We regularly read human-interest articles about this kind of love story, and no doubt about it, it appeals to the romantic side of our nature. So, if it's possible to fall in love in cyberspace, then isn't the presence of a signature odor irrelevant? To clarify this, let's look at the other side of the story.

Gayle, a young woman I interviewed, thought she'd found Mr. Right when she met an appealing man on the Internet. They "talked" back and forth via their computers for several months, and eventually they spoke on the phone and exchanged photographs. Some of their Internet communication became quite romantic, even erotic, and after about eight months, they arranged to meet in person. Needless to say, both expected an explosive reaction when they were face to face. Their story didn't make the news because as a romantic couple they were a flop! Gayle told me that they had no sexual chemistry—none. After a disappointing weekend, they parted friends; but it was a big letdown for both parties.

You may read about a man who finds the love of his life in a foreign country and learn that his first exposure to this woman was through a dating "catalog," which featured women who were looking for mates. He corresponded with a few women and eventually chose one with whom he seemed compatible. The two met and had an explosive romance. But you don't hear about all the couples whose correspondence courtships seemed to go well, but when they met in person, they weren't especially attracted—or one party was attracted, but the other wasn't. Such stories aren't particularly newsworthy.

Is Marriage All about Genomes and Pheromones?

Actually, if you look at mating and marriage across many cultures and over thousands of years, you can see that marriage is about much more than physical attraction. In the past, many cultures had some form of arranged marriage; many societies still prefer this form of mating. The conventional wisdom is that this system of choosing a mate is very successful, with success being defined as producing a stable and lasting bond.

Societies that still stringently regulate courtship and marriage would look at our culture, with our high divorce rate and unstable families, and claim that romantic love is not the ideal basis for marriage. No doubt, our culture has made a trade-off. We've allowed individuals to freely make their own choices about whom to marry, and divorce is not as stigmatized as it once was. The price we pay includes weaker family bonds—that's obvious. But we also experience greater financial insecurity because marriage is not viewed as an economic partnership to the extent it once was.

Arranged marriages may certainly appear stable, but we must also remember that divorce is not well tolerated in those cultures in which families are responsible for choosing a mate for their children. Happiness and stability are not the same thing. In addition, in some cultures in which individuals are expected to marry for social and economic status, it isn't unusual for a male to have a wife who bears his children and is his "public" partner and to also have a mistress. As long as these men, usually the wealthy and powerful, fulfill their duties to their wife and family and don't create a scandal, then their behavior is acceptable.

Human relationships are never simple, so though you're free to choose your own mate, you are influenced by your social group, including your family, friends, professional associates, and so forth. Nothing creates family upheaval more than a situation in which an adult child is said to be marrying the "wrong" person.

Societies and human motivations are too complex to reduce them to one factor. Just as tightly regulated marriage customs haven't prevented forbidden romantic love from stirring up trouble, our relative freedom hasn't resulted in courtship chaos. Accepting the existence of pheromones, substances that influence us without our conscious knowledge, simply sheds light on the chemistry of attraction and may lead to greater understanding of nature's ways.

4

What Is
Sexual Arousal?

SUBJECTIVELY, you understand what it is like to be sexually aroused, and it feels good, which is why we repeat the experience and keep our species going. Physiologically, however, arousal is a complex neurological process, in which sensory signals reach the brain and stimulate the changes in the body that we associate with sexual excitement.

Much of what we experience in sexual arousal are involuntary neurologic responses, much like our responses to food. If you inhale the aroma of baking bread, the message reaches the brain and you begin to salivate. If you're hungry, then your mind becomes preoccupied with eating. You didn't consciously order your salivary glands to activate; it was an involuntary response. Similarly, sexual stimuli cause spontaneous, involuntary physiological responses, and the stimuli may be self-generated, in that thinking—or dreaming—about sex can begin the arousal process. Even the expectation of seeing a lover can leave you in a pleasantly aroused state, simply because you are anticipating romance and sex.

A sexual encounter proceeds through five stages: arousal, also called the excitement stage; intense arousal or excitement; the plateau phase; the orgasmic phase; and resolution. For men, arousal is triggered primarily by visual and olfactory stimuli; for women, sexual excitement is primarily triggered by tactile and olfactory signals. These are generalizations, of course, and sexual arousal is ultimately an individual matter. Many women are aroused by visual stimuli such as erotic movies or the physical features of a particular man; men may become aroused by their favorite sensuous music or by the touch of their lover in the dark.

Each of our senses is controlled by a specific area of the brain. Sensory signals are processed in that area and are then projected to the septal nucleus, the name for the erection center of the brain in both men and women. A lover's touch, for example, is a message that travels through the spinal cord, is processed in the thalamus, and is projected to the erection center. At this point, the signal travels back down the spinal cord and results in the physiological changes we associate with arousal. Visual signals are processed in the occipital lobe before being projected to the erection center.

The Mystery of Erection

Because the male genitals are external, we've tended to think of erection as a male phenomenon, but the female clitoris is comprised of erectile tissue, too. In addition, the olfactory apparatus contains erectile tissue, which also becomes engorged during the arousal phase.

In men, an erection occurs when increased blood flow to the penis causes spongelike chambers to become engorged as the blood vessels of the penis expand. The scrotum also becomes congested, which results in a smoothing of the skin. As the arousal phase continues to the intense arousal phase, the scrotum flattens and the testes become raised. In some men, the nipples also become erect.

In women, arousal causes increased blood flow to the vagina, which expands the vaginal tissues and stimulates the release of lubricating fluids along the walls of the vagina. The cervix and uterus become elevated; this is a mechanism that allows deep penetration of the penis during intercourse, but without causing pain from deep thrusts striking the cervix. The labia majora (which translates from the Latin as "large lips")—the outer, fatty folds of the vulva—begin to flatten during arousal. The labia minora (the "small lips"), which are the hairless folds of skin just inside the labia majora, begin to lubricate. The clitoris is located inside the labia minora, and it becomes engorged during the arousal phase because of the increased blood flow to the region. The clitoris is covered by a membrane that is rich with nerve endings, which is why it is extremely sensitive to stimulation. Relative to the penis, it is a small organ, but its sole purpose is to provide sexual pleasure. Unlike the penis, it is not involved with passing urine from the bladder or with any other bodily function. Blood flow also increases to women's breasts during early and intense arousal stages, causing an increase in their size. The nipples also enlarge and become erect.

The sensory stimulation that takes men and women through the arousal and intense arousal stages leads to the plateau or pre-orgasmic phase, in which a high degree of sexual arousal occurs. Late in the excitement phase or early in the plateau stage, a phenomenon called the "sex flush" may occur in half to three quarters of women and among some men. This flush may look something like the rash that appears with the measles. It may spread over the breasts and chest, the abdomen and the buttocks, as well as the face. The flush may be more obvious on fair-skinned individuals than on those with darker skin, and its appearance, or lack of it, is not a measure of intensity of arousal.

During the arousal stages and the plateau phase, your breathing becomes rapid and heavier, your blood pressure increases, your heart beats faster and harder, many of your body's muscles are contracting, and you may begin to sweat. In the plateau stage, the penis becomes darker in color because of the engorgement, and the testes increase in size, anywhere from 50 to 100 percent. Fluid, which may contain sperm, may be released from the penis during the plateau phase.

Women's breast size may increase as much as 20 to 25 percent in women who have not breast-fed an infant. At the plateau stage, the inside of the vagina thickens because of the engorgement of the blood vessels, and the lubrication slows somewhat. Just prior to orgasm, the clitoris retracts, and increased engorgement of the labia takes place.

One of your most powerful drives is the urge to experience orgasm, or as it is also known, the climax. You experience the sensation of orgasm throughout your body, but the orgasmic "control" center is in the limbic portion of the brain. The electrical activity in the brain during orgasm is intense, with electrical surges that stimulate the ecstatic response we associate with this peak sexual experience. It follows that orgasms are intensely emotional experiences, so it isn't unusual for men and women to have highly emotional responses. Crying, laughing, or crying out are all normal and typical responses, although the range of individual variation is vast. A quiet response is not necessarily indicative of a lesser degree of pleasure or intensity than a loud, boisterous one.

Just before a man reaches orgasm, a two- to three-second period occurs in which he can do nothing to stop the orgasm. Seminal fluids have gathered at the base of the urethra, and the orgasm itself involves four or five strong muscle contractions, usually less than one second apart, that eject one or two teaspoons of fluid from the penis. This, of course, is called ejaculation. Male orgasm can be stimulated manually, orally, or through intercourse, all of which offer the opportunity to add variety to sexual experiences.

The resolution phase is simply the period of time after the orgasm, when the body returns to the pre-aroused state. The penis becomes flaccid, and gradually blood pressure and heart rate return to normal. This is a resting period for the penis, and depending on circumstances and age, a man may be able to achieve another erection in as little as fifteen minutes or may need as much time as up to a full day. Younger men and adolescents typically need less "recovery" time than older men, but stress and fatigue influence this, as do the presence of physical diseases and conditions and the overall state of a man's physical and emotional health.

The female orgasm involves vaginal contractions, anywhere from three to fifteen or so, and they are less than a second apart. Some women describe slow, long contractions, while others tend to have faster, shorter muscle spasms. Many women have both types of sensations, depending on the kind of stimulation they receive during a particular encounter. For many women, oral or manual stimulation of the clitoris results in the most intense orgasms, and some women prefer this to climaxing during intercourse. Individuals vary significantly, however, and no one method of achieving orgasm is more "average" and certainly not more "normal" than any other.

Women may not need a resolution period and can follow one orgasm with another or with several others—although this, too, varies among individual women. At some point, however, a resolution stage will occur, during which a woman's heart rate and blood pressure return to the pre-aroused state. Theoretically, a woman can be aroused again very quickly, but mood and desire will determine if this occurs.

In our cultural language, the resolution stage is referred to as the period of the "afterglow," a time that may also include "afterplay," which can mean anything from curling up for sleep to intimate or playful talk to heading to the kitchen for a snack. Some women say they feel energized after sex and ready for some good talk and a bowl of popcorn—or a shower or bath for two. While many men are game for some postsex activity, others become sleepy and may doze off in the middle of the playful talk! In general, however, sexual activity is usually relaxing, and the concerns of life are put aside for a while.

Where Do Odors Enter the Picture?

Odors are most influential during the arousal and intense arousal phases for both men and women. An odor can be as powerful as a stimulating picture or a tender touch in initiating the physiological changes associated with arousal.

As you become sexually excited, regardless of the initial sensory signal, you begin to hyperventilate, which coincides with nasal congestion. The congestion is caused by the engorgement of the erectile tissue in the nose. The congestion alters the airflow through the nostrils and obstructs the direct flow of air into the lungs. As I described in chapter 2, the congestion causes tiny "tornadoes" to form in the nose. These tornadoes increase olfactory ability, thereby allowing concentrated odorant substances to reach the olfactory bulb. This, in turn, influences the arousal response.

Odors can negatively affect sexual response, which is why you are probably concerned about your breath if you've eaten garlic or onions or other food that leaves a strong odor behind. You may also become self-conscious if you don't feel clean and fresh, and it is true that it is during the arousal phases that an unpleasant odor on your lover's body or in the environment can interfere with your sexual excitement. A truly bad odor can override the other sensory signals, pleasant though they may be.

Human mating tends to be a face-to-face activity, although obviously not exclusively so. As I mentioned earlier, when partners face each other, they have close access to the underarm area, which enhances the possibility of inhaling pheromonal substances and perhaps increasing arousal. Embracing and dancing also allow for release of pheromones, and these activities are frequently part of sexual foreplay.

Considering how important odors are to the first two stages of sexual arousal, it's easy to understand why the fragrances we use on our person and in the environment are the focus of multimillion-dollar "romance" industries.

The Complexity of Arousal

When we look at the physiological responses involved in sexuality, we might find that it all sounds quite clinical. Certainly, successful lovemaking depends on a range of physical reactions to stimuli, but men and women are more than a bundle of nerve endings and hormones. We bring our rational and emotional selves into the bedroom, as well as memories, both good and bad, of past relationships. We also bring our unique personality quirks and current conflicts or problems to our sexual experiences. Arousal is affected by fussing children, financial worries, the state of our careers, and our feelings about our partner at that time. We may desire sex with our lover when we're ecstatically happy or when we're sad and want the comfort of our lover's touch. We are drawn to sex to have romantic needs met, for physical release, or to feel emotionally close. Sex can be playful, passionate, affectionate, or tender and can involve complex combinations of motivations, moods, and settings. Sometimes we want to have sex simply because it's so much fun.

We're also individuals with differing drives and appetites for sex. Every now and then national surveys are published that tell us that we're either above or below the national norm in terms of how often we have sex and how many orgasms we have. Except in cases where couples have radically different sex drives, which can present a difficult problem to resolve, most couples would do better to forget quantity and concentrate on improving the quality of their sexual lives, which can carry over into areas of life outside the bedroom.

Ideally, couples will take time to understand each other's sexual appetites and preferences. Some individuals are aroused more quickly than others, and loving partners will adjust their lovemaking to accommodate these rhythms. They will also take care to create an odor environment that is pleasing and that enhances arousal.

In the next chapter, we'll discuss situations in which smells can cause confusion in sexual settings or even potentially damage relationships.

5

Smelling Bad/Smelling Good: What We Like— What We Avoid

MARY KAY isn't sure how to tell John that he doesn't smell appealing and that because of this, her sexual desire for him has lessened. She doesn't want to hurt his feelings, but on the other hand, being physically close to him has become unpleasant. Further, Mary Kay is resentful that John pays so little attention to his hygiene, and she's angry about his lack of consideration for her. Earlier in their marriage he always showered before bed and changed his clothing more often. Now he comes to bed in a ragged T-shirt that smells as bad as it looks. To make matters worse, he's started smoking cigars with his sports buddies, so he reeks of stale cigar smoke too. Doesn't he realize that he smells so bad that she's turned off?

Jim has a different problem. His lover, Jackie, thinks she smells bad, and she won't have sex with him until she takes a shower and scrubs herself with scented soap. Jim thinks it's nice to take a bath or shower together on occasion, but it isn't necessary, and besides, he likes Jackie's odor after they've been hiking in the woods or just sitting around watching a video at home after dinner. Sometimes he wants to make love on the couch, but Jackie won't hear of it—until she's taken her shower, which according to Jim, takes away the spontaneity. He wonders why Jackie's so self-conscious. In fact, her discomfort with what he calls "regular" smells is affecting their sex life, and he goes so far as to say that Jackie is "neurotic" about odors.

These two couples aren't particularly unusual, although both are probably on the extreme end of the spectrum when it comes to attention—or the lack of attention—to body odors. And the fact that relationship problems are developing isn't surprising. Mary Kay is disappointed, because her husband was a good lover before he became so sloppy. She doesn't want him "perfumed"; she just wants him clean. While Jim's case is a little different, it all comes down to individual and gender-related reaction to smells.

Recognizing Each Other's Odors

From animal studies we know that females recognize their young by smell;[1] studies of mothers and infants confirm that the same phenomenon exists in the human family.[2] We can also recognize our own odors. Two researchers asked one hundred volunteers to wear identical, freshly washed T-shirts continuously for twenty-four hours.[3] The T-shirts were then placed in a paper bag, and the participants were asked to identify the shirt they had worn. Seventy-five percent of the participants were able to correctly identify the shirts on the first try. Smokers were less frequently able to accurately identify the shirt they'd worn, as were women who expected their menstrual period to begin during the week they were tested. (As I've mentioned, a woman's sense of smell is least acute at the onset of menses.) In another study, researchers found that 90 percent of mothers and 30 percent of fathers were able to identify T-shirts worn by their children.[4]

So, a significant number of us can identify ourselves through odors, and mothers, considerably more than fathers, are able to identify their children. The inability of fathers to identify their offsprings' scent may have to do with the fact that, in general, men's sense of smell isn't as good as women's. Another reason could be simply that fathers tend to spend less time with their children and, therefore, are less familiar with their odor.[5]

Individuals are also able to distinguish between male and female odors, and in general, the odors identified as male are considered stronger and not particularly pleasant.[6] Female odors are generally considered "lighter," or less intense, and are more often rated as pleasant. What is important is that both men and women rated male and female odors this way. In other words, women and men both found male odors less pleasant than female odors. What does this mean for social organization and relationships? If men don't like each other's odors, this suggests men will literally "keep their distance" from each other. In social or business situations, men tend to stand farther apart and have less physical contact with each other. Even the most acceptable touch between men, the handshake, tends to maintain distance. Women, on the other hand, are culturally free to touch each other frequently, and they tend to maintain close physical distance when they talk and often exchange casual hugs as signs of affection—they don't even need to be extremely close friends to do so.

In our culture, we have explained this difference in the context of *touch*. We say that men are inhibited from physical and even emotional closeness because of the expectations we have about what is "manly" behavior. But we could suggest that the differences in social behavior between men and women were originally initiated by our sense of smell. Perhaps in the ancient past, men needed to be distant from each other because of the competitive need to survive and not only "win" female mates, but also to gain social and economic status within the group. This may seem outdated now, but many men do seem to view one another as competitors, even if they are also friends.

What Men Can Learn about Women

Women are influenced by their perceptions of themselves in very specific ways. This fact is not surprising because women, far more than men, are judged by their appearance, and the pressure on them to look and smell good manifests itself in many ways, including the sexual arena. This may sound humorous, but if a man looks forward to a great evening of dinner, perhaps a movie, and lovemaking, he'd be better off reversing the order. Even though, as we'll see in chapter 8, food and sex go together, many women are so self-conscious about their bodies that they prefer to eat *after* they've had sex. They believe they look better before eating than they do after a meal, because right after eating, they're concerned about appearing bloated.

Before men—and perhaps women, too—conclude that this is ridiculous and a bunch of hogwash and ask how women could believe such nonsense, try to remember why women in our culture might have these thoughts. Women literally are bombarded with messages that tell them they're too fat, even when they're obviously not. The ideal woman, according to advertisers, is tall and thin with a flat abdomen. Women may see their partners glancing at shapely young women on the street, and they don't believe they measure up by comparison. Most women are not tall, and they aren't as thin as fashion models, so their self-perception leads them to think they aren't "good enough." Women with good body images have often had to fight for them, so to speak, and have avoided the heavy pressure of advertising.

Some women who seem overly modest in the bedroom and don't like their lovers to see them completely naked may not be particularly modest; what they are is self-conscious about their bodies. Men can— and I'm sure often do—reassure women about their bodies, but as long as our society is obsessed with a false idea about the "perfect" weight and body, women's self-consciousness is likely to continue on its present course. I don't want to overstate this, however. I realize that many women are comfortable with themselves and are not self-critical. However, for numerous couples, body image is a problem, which can become connected to other things—especially body odors.

In the typical couple, the woman is younger than the man and inherently has a better sense of smell. As I've stated, a woman's olfactory acuity fluctuates throughout the month, generally reaching its peak in mid-cycle. As the couple grows older, the man's sense of smell begins to diminish first, and it wasn't as sharp as hers in the first place. Women often notice odors that men don't, which makes them not only self-conscious about their own odors, but very aware of their partners' odors, too. Men might say that they don't notice that the house smells "off" or that their sweaty workout clothes are stinking up the laundry basket. However, women are likely to notice these things, and some say that men try to deny that they or the dirty clothes smell bad! We can conclude that, contrary to society's definition of the ideal match, in which the man is older than the women, nature's ideal "olfactory match" is for the woman to be many years older than the man.

In surveys of women about their odor preferences, the clean smells seem to win out. When I interviewed a group of young married women, they said they associated a soapy smell, or even a bubble bath scent, with cleanliness, which was a sexy smell for them. They also liked the odor of coconut suntan lotion because it reminded them of pictures of suntanned men who looked sexually attractive. The smell of suntan lotions probably also brings thoughts of the ocean and the beach, which are associated with clean, fresh smells. Even the slightly salty, fishy smell of the ocean is a positive odor because it is associated with the beauty of the beach. The odor of vanilla is also considered a clean smell, which probably explains why this is such a popular artificial scent used in candles and sachets and other odorized products for the home.

In a survey conducted in Wales, 70 percent of women said that men who wear cologne or aftershave are sexier than those who don't.

According to conventional wisdom, women wear perfumes and colognes to attract men, but the overall results of this survey suggested that women are more influenced by male odors than men are by women's scents. (Studies have found that women wear artificial scents because they like them and feel better wearing them. Sexual attraction is a secondary issue.)

A story is told about Somerset Maugham, who was puzzled that the novelist H. G. Wells had the ability to attract many women. Wells was overweight and considered rather homely, but he had no trouble finding lovers. When Maugham asked one of Wells's lovers why he was so successful with the opposite sex, she told him that Wells smelled of honey. I'm not recommending that men dab honey behind their ears, but men should note that if a woman says that a man smells of honey, it means that she thinks he smells sweet—pleasant and attractive.

Men might also note that some of the married women I interviewed told me they like to share odors, meaning that using the same cologne as their lover makes them feel closer to the man. On a primal level, a shared odor makes women feel a sense of oneness and acceptance by their partners, and hence, they feel more secure. This explains why women like to wear their partners' shirts or robes. If they like the scent of the man, then they want it close to them.

Some data suggest that when a couple live together, they begin to share skin bacteria.[7] While these organisms act differently on the sexes, women are able to identify their partners by their odor in part because of the familiarity with their smell.

Women also tend to be more self-conscious about their own mouth odor than men are, and because their sense of smell is usually better, they are sensitive to their partner's mouth odor, too. Some women won't eat certain foods if they believe it will affect their breath. These women are willing to pass up a glass of wine or a plate of garlic-spiced pasta or even a cup of coffee because they don't want to have offensive mouth odor. To men, this might seem extreme, but women are more self-conscious about odors.

In most areas involving odors, women are self-referring, meaning that they are conscious of their own odors and make efforts—sometimes to an extreme—to be sure that their odors are pleasing. Jackie really doesn't believe Jim when he says he likes her natural smell. In truth, what seems to him like a faint odor about her person after they've walked in the woods or taken a bike ride seems like a strong odor to her. She feels sweaty, and therefore, she thinks she must smell bad. He thinks the slightly sweaty odor is alluring. But, given the differing olfactory acuity between men and women, what seems like an overwhelming odor to Jackie may be just a faint and pleasant aroma to Jim.

When Self-Consciousness Becomes Serious

For the most part self-consciousness about odors in our culture is part of the way we're raised. Our society is very "cleaned up." We bathe more than any other culture in history, and our standards for personal cleanliness are very high. We spend millions of dollars on mouthwash, and advertising has made us very conscious of bad breath as a social problem. Women and men alike are conscious of our high standards, but women tend to be more aware of them, and personal hygiene is often an important part of their daily routine.

Women also tend to be more sensitive to rejection; so, for some women, a belief that they don't smell pleasing leads to a belief that they'll be rejected. Many studies have confirmed that women are more sensitive than men to the feelings of others. If a woman feels rejected, she may, among other things, question her hygiene and wonder if the negative response to her was based on not having a pleasant odor.

Most of the time these concerns fall into a normal range. But in certain cases, usually in conjunction with other mental disorders,

self-consciousness can become pathological. For example, some patho-
logically depressed people will become convinced that their bodies are
rotting from the inside out. They believe they are exuding a powerful
stench, and no amount of reassurance will convince them otherwise.
Some emotionally disturbed women begin to believe that their vaginal odor is so strong that they can't go out in public. Obviously, we're
talking about a small number of people, but concerns about body odor
are common among the mentally ill. And among depressed people,
more women than men manifest this extreme concern about their
own body odors.

Sometimes, particularly among schizophrenics, having a bad odor
is perceived as valuable and positive. We've all seen "street" people who
appear to be living in a different world. They may be very dirty and
seem not to even notice others around them. In some cases, these men
and women are using their body odors to distance themselves from
other people. Smelling bad keeps the rest of us from invading their
private space. (This is entirely different from the case of homeless
people who attempt to stay socially acceptable but don't have access
to the facilities they need to be clean. That's a sociopolitical problem,
not a psychological one.)

For the most part, the concerns women have about the way they
or their clothing or homes smell are not signs of more serious problems. Good communication can usually make a woman feel more at
ease with her body and its odors. What Jackie doesn't understand and
Jim hasn't been able to communicate is that her individual odor—her
odor signature—is a powerful attractant.

Jackie is beginning to realize that her self-consciousness is a problem that stems from childhood. Her mother was, as Jackie put it, "a
cleanliness nut." Jackie was told that smelling bad meant she was bad;
to be dirty was to be disgusting. So even the grime all children pick
up in normal playing had to be scrubbed away before she could be

with other people. It's no wonder that Jackie carried this attitude into adult relationships.

If you find yourself extremely self-conscious about personal cleanliness and odors, you might look into your background and see if you were taught that body odors are inherently unpleasant. If your partner seems overly concerned about hygiene, realize that the underlying issue is a fear of being rejected.

What Women Can Learn from Men

In Mary Kay and John's case, cited at the beginning of this chapter, it's likely that John doesn't realize that he smells bad. First, his sense of smell isn't as good as Mary Kay's, so he doesn't believe his T-shirt stinks. In fact, men tend to see their own odors as positive. At best, they don't think as much about hygiene as women because they aren't judged as harshly about their bodies and physical appearance. Because men are judged by different standards (for example, financial and career success), they tend to have positive self-images and inflated egos, and so they conclude that since they are "good," what they create must also be good. Therefore, the odors they produce are good, or at least don't smell bad.

Some years ago, I saw a published letter to one of the popular advice columnists from a woman who said her husband's hygiene had become terrible, and she didn't know how to tell him that she couldn't have sex with him because he smelled so bad. (Sounds just like Mary Kay's concern.) She was avoiding any physical contact with him, and she missed their regular sex life. I no longer have the newspaper column, but as I recall, the advice expert suggested that she lure her husband into the shower with her and make cleaning him up part of the sexual foreplay.

This is certainly one way to handle this problem, but what happens a few days later? If he doesn't take the hint, what is she to do? Women tend to be afraid of hurting other people's feelings, particularly those of people close to them. But if life is to return to normal, or if the source of the problem uncovered, then Mary Kay, and other women in the same situation, must call attention to the change in their partners' personal habits. Mary Kay can tell John how much she likes the smell of the soap on his body, or she can talk about how much a particular aftershave arouses her. If John still doesn't respond, she may consider that the problem goes deeper and may be a sign of something more serious.

Losing interest in self-grooming is one sign of depression or illness. We certainly see this in animals. You know your cat is ill when he or she is no longer fastidious about grooming. Well, the same is true for humans. It is rare that someone is so egocentric to believe that his unwashed body actually smells just fine. Stress and overwork can also lead to a gradual loss of interest in personal hygiene. It's also possible that a refusal to deal with the problem is a way of creating distance from a partner and a relationship. Women have often said they know when their husbands are being unfaithful because the men are taking great care about how they look—and smell—when they leave the house but are sloppy and unkempt at home.

So, while using subtle measures to let a man know that his hygiene is lacking may work in certain situations, if the man refuses to deal with the problem, then I suggest looking for another cause.

That Scent Is Making Me Furious

Our reactions to odors occur within a social context. Let's say that your mother always wore a particular perfume, but your relationship with your mother is rocky at best. It isn't likely that you will be attracted to a woman who wears the same scent. Along the same lines, even if you have a great relationship with your mother, you may not have romantic feelings about a woman who is wearing your mother's scent. This is related to the incest taboo, of course, and for most men, reminders of their mother will interfere with sexual arousal with their lovers.

Other smells may have similar effects. While the smell of baking bread, for example, makes most of us relaxed, and we enjoy the pleasant odor, the context may not make us think about romance and sex—we might think about grandma's kitchen instead. This isn't true for everyone, of course, but if you find that a scent is more of a distraction or even a turn-off, rather than a turn-on, even when it's pleasant, then check the context. You may find that odors are reminiscent of situations that aren't compatible with erotic thoughts and activity.

One woman never dated men who smoked because she associated that smell with her father, a man who was emotionally detached and ultimately committed suicide when she was a teenager. It's not difficult to see that this smell took her back to a troubled time in her life. Men who had the same cigarette odor on their person and clothes were actually repulsive to her. On the other hand, the same smell of tobacco products may be a pleasant odor to a woman who has positive feelings about her relationship with her father, who smoked. Obviously, I would never recommend that anyone smoke, but some people like the smell of pipe smoke, for example, because they have pleasant childhood memories of the smell of pipe tobacco filling their living rooms, and they may associate this with family closeness.

A more typical situation may be subtler, especially for women. Women tend to associate their sexual experiences with a total mood. If, for example, a woman has an argument with a man, the odors in the environment may become associated with that unpleasant experience. Those odors may influence the mood on other occasions.

If It Smells Bad, It Is Bad

For a variety of reasons, our culture has linked bad smells with evil—and evildoers. Witches were said to smell bad, adding proof of their evil natures. Conversely, certain saints are said to have given off sweet odors when they died, further proof of their inherent goodness. Some people think that street people must be bad or dangerous because they smell bad. In Nazi Germany, Jews and other victims of Hitler's brutal regime were said to smell bad; in our country's racist social structure, it was once claimed that African-Americans smelled bad, which helped justify racial segregation. These claims were part of reinforcing prejudice and hate, and it's alarming to realize how well these tactics worked!

The body itself has been viewed with suspicion in our culture, and some people are raised to believe the body is bad, so the body's odors must be bad, too. This, in turn, has resulted in our tendency to cover the natural odors with other scents. We use more deodorants and mouthwashes and hygiene products than any other culture. I'm not saying this is wrong. It is an undeniable trend that's not going to stop, and most of us want to smell good, however that's defined. This emphasis on smelling good is also part of our business culture, not just our social and romantic lives.

Our cultural trends are not a matter of "bad" and "good." Odor preferences among individuals and the two sexes aren't a matter of right or wrong, either. Adjusting to each other's sensibilities about odors is an important part of an ongoing sexual relationship. Ultimately, we must enjoy our partner's scent, or we will not enjoy a romantic relationship. In fact, one study suggested that disliking the smell of the other was a predictive factor for the failure of the relationship. Our odor likes and dislikes no doubt influence the kind of sexual practices we like, which affects our compatibility as couples, too.

Minty-Fresh Breath

For some reason, a person's breath is one of the most delicate subjects to discuss. Regardless of the reasons, however, it is an important part of sexual attraction, and when a person's breath is perceived as "bad," then social, business, and sexual relations suffer. The odor of the breath varies, of course, and is influenced by diet, tobacco use, and oral hygiene, and in women, perhaps even by the pheromones produced during the menstrual cycle.

Halitosis, which is the medical word for bad breath, is sometimes caused by illness, such as an infection in the mouth or sinuses. It can also be caused by tooth decay or gum disease. If bad breath is a persistent problem, and attempts to correct it don't alter the situation, then I recommend a medical and/or dental checkup. In most cases, however, the issue is resolved by improving oral hygiene.

In our culture, we make attempts to either neutralize the breath or artificially make it smell pleasant. In 1997, our Smell and Taste Treatment and Research Foundation surveyed one hundred men and one hundred women, and we found that 80 percent of women liked the odor of mint on a man's breath. A smaller number favored chocolate or fruit. No women actually liked the smell of alcohol on a man's breath. This was true for both single and married women.

Our survey found that forty-two men preferred alcohol on a woman's breath, while thirty liked mint, and eight preferred chocolate. It's interesting to note that all forty-two men who liked the odor of alcohol were *single*; not one married man preferred it. When we asked the single men about this, they said they associated the odor of alcohol on a woman's breath with her sexual availability, and that, perhaps, some alcohol would make her more receptive to sex.

Our foundation once conducted a study of olfactory ability and alcohol intake.[8] The study subjects were State of Illinois employees, supervised by a state trooper, because the study involved drinking significant amounts of alcohol. The study volunteers were told to drink the alcoholic beverage of their choice (wine, vodka, beer, piña coladas, and so forth), and as they became progressively drunk (established by measuring blood alcohol levels), their olfactory ability diminished, which is not good for sexual adventures. We could say that alcohol temporarily "damages" olfactory ability.

When men believe women may be more receptive to sex when the women have been drinking, they should consider what animal studies tell us. In males of various species, damaging the olfactory apparatus often diminishes performance; in female rats, for example, damaging the olfactory center reduces the female's ability to discern appropriate mates, and she mates with any available male rat. So, if you're a single man, and a woman who has been drinking heavily agrees to have sex with you, don't be excessively flattered. Consider that her olfactory ability is probably impaired, and any mate would do! Some individuals of both sexes have had the experience of waking up in the morning and being forced to wonder what in the world they were thinking upon seeing a stranger or an otherwise inappropriate partner in their bed. They blame the alcohol, which certainly did diminish their judgment, but perhaps they should also consider that their sense of smell steadily decreased throughout the evening and interfered with the process of making a sound choice.

With the exception of some single men, most men and women probably prefer the odor of mint because of its association with toothpaste and mouthwashes, as well as with breath-freshening candy. Chocolate also has a sweet odor and has long been associated with romance; chocolate and mint are even combined in candy and ice cream.

As I've said, women tend to be self-conscious about mouth odors, sometimes excessively so, because they tend to be self-critical. But, men may need to think more about oral hygiene if they would like a receptive partner. Breath that smells of beer and whiskey isn't welcome in the bedroom.

Oral Sex and Odors

Oral sex is a complex issue in our society, mainly because taboos against it exist that have even made their way to our legal system. In many states, oral sex is covered under the sodomy statutes, meant to outlaw homosexual behavior. Nowadays, these laws are only sporadically enforced, and I'm confident that eventually they will be removed from the books. Be that as it may, our society delivers mixed messages about oral sex. In our own bedrooms it may become an issue because of early influences that may have biased us against it. In any case, for some, the odors are part of the attraction of oral sex, and for others, odors are the main detraction.

According to one survey, 75 percent of men would like oral sex more often, and of that group, one third said it was the most pleasurable part of sex for them. On the other hand, about 40 percent of women surveyed said they didn't want to try anything new in bed, and 20 percent said they'd refused an offer or a request for oral sex. Some surveys suggest that women who are most resistant to this practice—either giving or receiving—were somewhat older than women who are more receptive, which may reflect the fact that they were raised with strict religious and cultural prohibitions against it. These prohibitions were more common in the past than in today's society. Age is not a determining factor, however, for either men or women.

We don't know for sure if discomfort with oral sex is unique to our society, but we do know that the *Kama Sutra* and *The Perfumed Garden*, sex manuals written in much earlier times, included advice about performing and receiving oral sex for both men and women. ("Cunnilingus" is the term used for performing oral sex on women;

"fellatio" is the word used for performing oral sex on men.) Most modern sex manuals discuss this issue, too, and in *The New Joy of Sex*, the author, Alex Comfort, calls oral sex "mouth music."

Those who resist the idea of oral sex usually have been conditioned to believe that the genitals are dirty, and in extreme cases, they don't want to even touch the genitals of their lover. If cleanliness is the issue, the mouth actually has far more active bacteria than do the genitals. Many women say they are self-conscious about their vaginal odors and the taste of vaginal secretions. Many women fear that these odors and tastes will offend their partner. This may be a sign of our times, in that natural odors don't have the same "status" that they once did.

In one of Napoleon's letters to his wife, Josephine, he tells her that since he'll be home in two weeks, he doesn't want her to wash. Just thinking about her sexual smells was exciting to him. But the scent of violets was Josephine's preferred perfume fragrance, and we know that she used it on her body and clothing—washed or not. The decision isn't between going completely natural or covering the body's odors with external fragrances. Most people like a combination of odors; this isn't an either-or issue.

For many men, performing oral sex is exciting in part because it's a way to please their partners. Oral sex provides the opportunity for the clitoral stimulation necessary for many women to reach orgasm, and the vaginal odors are sexually arousing to most men. If odors are the main concern for the woman, then sharing a bath or shower or a massage with aromatic oils as a prelude to sex may be relaxing and provide the woman with the reassurance she needs to enjoy oral sex.

Vaginal odors also change during phases of the menstrual cycle, and the odor becomes less intense as ovulation approaches. In one study of changes in vaginal odors, the researchers reported that both men and women found the odor less pleasant when it was less intense. However, this particular study introduced the odors in a laboratory setting, not in a sexual context, and overall, the odors were not particularly arousing to either sex. Our perception of any smell depends on the setting, the environment, in which we detect it.

It isn't surprising that vaginal odors are altered during menstrual cycles, because vaginal secretions are produced by numerous compounds, perhaps thirty or more, including those released by the sebaceous glands surrounding the vulva and the internal glands in the duct of the ovaries and the endometrium, the lining of the uterus. These substances can be affected by bacteria, which consequently alter their odors. Vaginal secretions, by their very nature, are "unstable" biochemical compounds, in that their production and composition are subject to influence by many other physiological processes.

When a woman's vaginal odor is actually quite strong continually, then I recommend she be checked for a vaginal infection. These infections are quite common, and because bacteria change the chemistry of the secretions, their odors also change. (Trichomoniasis, usually a sexually transmitted infection, produces particularly strong-smelling vaginal secretions, and in fact, a strong, unpleasant vaginal odor is one of the symptoms of the condition. The vaginal fluid may become yellowish in color, and the vaginal and vulvar regions may itch. In this case, the partner should be examined and treated as well; although men do not necessarily become symptomatic, they can reinfect their partner if the condition is left untreated.) Unfortunately, some women believe their odors are strong and offensive all the time, so they resist oral sex altogether.

Some women object to male genital odors and, therefore, object to performing fellatio. Women who do enjoy these odors and like performing oral sex may not like the taste of the ejaculate—the prostatic and seminal fluid. The chemistry of these fluids, just like vaginal secretions, changes just as the chemistry of our sweat or urine changes. Our diets and habits affect these odors and tastes, too. For example, some women say that the semen of men who smoke has a less pleasant taste than that of men who don't smoke.

Thus far, the discussion has centered on men who enjoy oral sex but who have partners who don't. Obviously, this is not always the case. Some men don't like either cunnilingus or fellatio, and their female partners are disappointed. The same advice about handling these differences applies to both sexes. Communicate with each other and learn what the other person's objections are and go from there. If odors are the issue, then use ways to remedy the situation by adding pleasant odors to the sexual encounter.

When the objections are related to the *idea* of oral sex, then perfumes and massage oils are not going to solve the dilemma. We are all the products of our upbringing, and for some people, reservations about various sexual practices are psychological issues and not related to hygiene. Of course, no universal agreement exists about sexual practices, which vary not only from culture to culture, but among individuals in the same society. Objections to particular sexual acts are a problem only if they create difficulties in a relationship. Otherwise, they are no cause for concern. We all want to please our partners in the context of loving relationships, and for a variety of reasons, some practices are unappealing or even repulsive to some people. What we often do, however, is take a rejection of a sexual practice as a personal rejection. It's no wonder that sexual relationships are often complicated and not as easy as they look on the silver screen. We bring a host of attitudes to bed with us, and change is never easy.

What Is That Odor Doing in Here?

In "The Invalid's Story," Mark Twain tells of a man who jumped aboard a boxcar and found himself huddled in the dark next to what he thought was a coffin, because of its terrible smell. The man was so traumatized by being trapped with the stench of a dead person that, finally, unable to take it any longer, he jumped off the train and landed in a snowbank. Later, the man was found, gravely ill with pneumonia, and was taken to the hospital. After telling others about what had happened to him, they investigated and discovered that no coffin and no dead person had been on that train. But a big crate of cheese had been, and the odor had permeated the boxcar. They went to tell the ill man what they'd found, but it was too late. He had died, and the indirect cause of death was a crate of strong-smelling cheese.

This story illustrates how important context is in determining our response to odors. The odor of garlic in a pasta sauce smells great, but how would you react to a garlic-scented candle? Responses to particular odors are individual, and we are conditioned to like or dislike different smells in context.

Citrus smells are generally rated fresh and clean, and many commercial products are odorized with lemon, for example. If you associate a lemon scent with dishwashing soap, then you might not respond positively to a lemon-scented room spray for your bedroom. The scent of orange blossom and orange essence, however, is an odor linked with romance and sexuality, and so far, we don't see orange-scented household cleaners on the market. If that were to happen, and these cleaners were to become popular, then we'd probably find far less orange fragrance in the bedroom.

If your lover responds negatively to an odor in your bedroom or on your clothing, then consider the possibility that the context isn't compatible. An odor sensuous to you might be a turn-off for your partner. Such a difference can be subtle, so be sure you both agree on the scents you use in a romantic, sexual context.

Talking It Through

The best way to solve difficulties in sexual relationships is to improve communication between you and your partner. Whether you're dealing with hygiene issues or specific likes and dislikes in bed, talking about these concerns leads to greater understanding. You may find that the problems are truly small. A man may not like a particular perfume a woman wears, or he may find her scented shampoo or even the fragrance of the candles unpleasant. A woman may enjoy performing fellatio when her partner is fresh from a shower, but dislike the practice at other times. Any possible combination of likes and dislikes is possible.

The best way, the only way, to solve these relatively minor difficulties is to speak up. Most sex therapists advise talking about sexual preferences and potential or actual problems outside the bedroom, so that emotionally charged issues don't interfere with the romantic atmosphere conducive to pleasurable lovemaking. I agree with these therapists when it comes to talking about odors and sex. We know odors can enhance our sexual lives, but they can interfere with our pleasure, too, so don't let concerns and resentments build. Use the information in this book as the impetus to start talking about sex and the fragrances of romance.

6

Smells Men Find
Irresistible

THE LINK between a physiological response to odors and sexuality is clear and direct. You may notice that your nose becomes stuffed up when you're sexually aroused. Of course, on some occasions it may be more noticeable than others, but it happens as a result of a built-in physiological response. You may believe you're different in some way, because having a stuffy feeling in your nose isn't something we generally talk about, and most books describing sexual arousal and lovemaking techniques don't mention it. But you usually begin breathing through your mouth when you are becoming sexually excited, and when you hear the expression "coming up for air" in relation to passionate kissing, it's usually because you're finding it difficult to breathe normally.

In chapter 2, I described the pathway of an odor molecule, and I mentioned the tiny tornadoes that form in the nose. When these air currents form, less air goes straight to the lungs, which in turn allows more odor molecules to bind to the olfactory epithelium at the top of the nose and, hence, be registered as a smell. This means that the erogenous nose is able to perceive greater concentrations of pheromones and other odors that increase sexual arousal. This mechanism is another of nature's ruses to keep the species going. As various fragrances, and the pheromones we aren't consciously aware of, reach the brain, the more sexually excited we become, unless some other overriding factor interrupts this process. But, we wondered, are all odors created equal in their ability to arouse us, or are some scents better than others?

The Study Develops

At the Smell and Taste Research Foundation, we test people who have lost their sense of smell through a variety of causes, but most commonly due to head trauma as a result of automobile accidents. In the course of this testing, we noted that about 25 percent of those whose ability to smell is significantly diminished also experienced sexual dysfunction. Animal studies certainly link olfaction and male sexual function. For example, if we destroy or remove the olfactory bulb in a male mouse, he will lose interest in mating.[1] Many animal studies support the link between sexual behavior and olfaction in both sexes.[2] This phenomenon led us to further explore the connection between our sense of smell and sexual functioning.

I hypothesized that if loss of ability to smell can lead to loss of sexual functioning, then perhaps odors are necessary for sexual arousal. If this is so, and we exposed those with a normal sense of smell to particular odors, then sexual arousal would be enhanced. Alternatively, if we provided other odors, perhaps they would inhibit sexual arousal. Further, if odors that increased arousal could be found, this discovery would represent a potential treatment for impotence. In addition, if some odors were found to inhibit arousal, then they could be used as part of stimulus deconditioning in sex offenders.

In our first study, conducted in 1994, the twenty-five volunteers were male medical students. The traditional way to study male arousal is to measure penile blood flow, which is accomplished by attaching a small blood pressure cuff to the penis. This is the method used to measure impotence and other sexual dysfunction. We also attached a blood pressure cuff to the arm to be sure that the odors we introduced didn't cause generalized changes in blood pressure.

The study was double-blind and randomized, meaning that neither the research team nor the individual subjects knew which odors were used or the order in which they were introduced. Each subject wore a surgical mask that had been infused with one odor. Each odorized mask was worn in place for one minute, which ensured that we eliminated what is known as the "Hawthorne effect." The Hawthorne effect established that any novel stimulus will induce a response of some kind. Keeping the masks on a full minute meant that changes brought about by the novelty had time to wear off and, therefore, wouldn't skew our results. After one minute, the changes in penile blood flow were measured for each odor and were compared to any changes in the blood pressure readings taken at the arm. Three-minute "wash out" intervals came between the introduction of odors, during which time no mask was worn, allowing the nose and the penis to return to baseline.

The group of odors we tested included a variety of common floral scents and perfumes, and for the purposes of having what is known as a control, we added in the scent of baked cinnamon buns. We didn't expect that odor to have any effect, but it would serve as a scent whose "performance" we could compare with other odors, which were presumably associated with sexual or romantic settings. Much to our surprise, the cinnamon buns caused greater changes in penile blood flow than any other odor!

Of course, we didn't know what our results meant. Did our study "prove" that medical students are always hungry? It could have meant just that. We also didn't know if the results would be the same among men in the general population and across a wider age range, so we planned another study, which took place in 1995.

We recruited volunteers during an afternoon appearance on a popular Chicago radio station, known as "The Loop." We wanted a wide age range represented, and the thirty-one males who enrolled in the study ranged from ages eighteen to sixty-four. The majority of subjects had olfactory ability that fell within the normal range. Some subjects had diminished ability to smell. The olfactory ability of the subjects was tested using the standard procedures and tests we use when people come to our office for treatment of smell and taste disorders. For example, we use a forty-question "scratch'n' sniff" test, in which the person is asked to identify many common odors, such as leather, bubble gum, and natural gas.

For this study, we isolated thirty scents, which represented many "categories" of smells. The total number of tested odors was forty-six because we combined some odors. The protocol was the same as that used for the first study, and all the subjects were exposed to all forty-six scents.

While the majority of our study subjects were heterosexual, a few were homosexual, and most of the men had a regular sex partner. We also questioned the men about the kinds of odors they liked, and all told us that they'd never had an odor-induced erection. In other words, they weren't able to isolate any particular odor that they associated with arousal, although they said they enjoyed various perfumes and other odors they considered part of romance.

More than half of the men had experienced odor-induced memory recall, meaning that certain smells triggered an incident or feeling state from the past. Like most men, they experienced the phenomenon of morning erection, an event that begins in puberty and continues throughout most of adult life. To summarize, the men we recruited were fairly typical American men, much like you or your relatives and friends.

We're often asked how the men could become sexually excited

when they're hooked up to machines in a laboratory setting. But both the response to an odor and penile blood flow are essentially involuntary responses. Blood is flowing through the body all the time, and the changes in blood flow to the penis can be changed based on a variety of stimuli. The stimuli in this case were odors, including perfumes, food smells, and floral scents.

Were They Aroused or Were They Hungry?

This particular study brought great interest from the press. The subject was amusing in itself, because Americans tend to find sexual topics funny and even somewhat embarrassing. So, can you imagine how the press reacted when we announced that the smell that caused the greatest measurable arousal in men was a lavender and pumpkin pie mixture? This odor combination increased penile blood flow by an average of 40 percent. A close contender was—you guessed it—cinnamon buns! Of course, the obvious joke was that when our grandmothers said that a way to a man's *heart* was through his stomach, they were apparently speaking symbolically—and politely.

Various newspaper columnists began calling our office and asking more questions about our study. Soon some were jokingly advising women to buy "eau de pumpkin" perfume and makeup scented—and flavored—with "essence of cinnamon roll." Might as well throw away those expensive scents with seductive names and start stocking up on frozen pumpkin pies! Jay Leno made jokes about this study on his show for three straight evenings; Oprah Winfrey teasingly tried to steal away our "active formula," as did Joan Lunden when we appeared on *Good Morning America* to talk about our study.

One thing the press didn't report that I find amusing is that the man who showed the strongest response to the lavender and pumpkin pie combination was asleep throughout the study! Remember, I explained that the blood flow response is involuntary. The men didn't need to be thinking sexual thoughts when the odor was introduced, and it wasn't even necessary to be awake. The brain, as the saying goes, has a mind of its own.

Other "winning" odors in our study included a combination of licorice and doughnuts, which increased penile blood flow by a median of 31.5 percent. A pumpkin pie and doughnut combination was also a winner, with a 20-percent penile blood flow increase recorded.

What about odor losers? Thanksgiving might not be the same without pumpkin pie, but apparently if love is on your mind, don't inhale the cranberry sauce. Cranberry scent was the biggest loser in our study, averaging only a 2-percent increase in blood flow. For the men in our study, chocolate is not a powerful odor, and it came in near the bottom of the list, as opposed to licorice, which "performed" very well. Some of the odors might not seem particularly logical choices, but we included them because we wanted different categories of odors. Such was the case with baby powder, which came in near the bottom of the list.

For some men, buttered popcorn showed a moderate response, as did orange scent and musk. Licorice and cola in combination increased penile blood flow more than either odor alone. In fact, the combination odors were found to be the strongest. This isn't too surprising, because food odors tend to occur in groups.

We made some other interesting findings. For example, the older participants had a stronger response to the odor of vanilla than men in the younger age groups. The men who reported the most satisfaction with their sexual lives had greater engorgement in response to the odor of strawberry. Those who had sexual intercourse with greatest frequency responded the most strongly to the odor of lavender, as well as to Oriental spice and cola.

Of the group of odors we tested, lavender appears most often in the folklore about sexual scents. Vanilla is sometimes mentioned, but in general, food odors are not as commonly associated with sexuality as they probably should be.

(As an aside, no odor diminished penile blood flow, which probably comes as no surprise to female readers. Even cranberry stimulated a response, weak as it was. More research will determine if we can find an odor that could perhaps be used as part of aversion treatment for sexual offenders. This was a peripheral issue in our study, and we will continue to look into it.)

So, Why Do Men Respond So Strongly to Food Odors?

We had predicted that men would respond to odors considered positive, and the results supported our original hypothesis. It's possible that these odors induced a Pavlovian conditioned response; in a primitive way, the smell of food reminded the men of their lovers, and stimulated the associated physiological response, that is, enhanced penile blood flow. We know that the odor of baked goods evokes olfactory-induced recall in most people, which is much like the Pavlovian response to stimuli. Whiff some baking bread, and you might think about a bakery in your childhood neighborhood or the smell of baking in your kitchen. These recalled events are generally pleasant and subtly alter moods. Thus, it's possible that the food odors put the men in a positive mood state, leaving them open to a range of thoughts, some of them sexual.

There are many reasons for the positive response to food odors, even a positive sexual response, but the truth is, we don't know for certain why these responses occurred. One possibility is that many of the odors had a relaxing effect. For example, we know that green apple reduces anxiety, as does lavender.[3] These odors increase alpha waves in the back of the brain, a response that is associated with relaxation.

Sexual response is usually the most positive when men are in an alert, awake state, but are also relaxed. As we all know, stress, anxiety, worry, and tension are not aphrodisiacs.

When the men were relaxed, and any prior anxiety they might have experienced in connection with being in a study was reduced, their inhibitions might have been removed, and thus, blood flow to the penis increased. Alternatively, we know that some odors increase beta waves in the front of the brain, which is associated with a more alert state.[4] Jasmine is one of these odors, and it is also associated with sensuality and sex in some of the popular folklore and aromatherapy. Peppermint, an odor we tested, also has this effect. In fact, alarm clocks that release the odor of peppermint have been developed in Japan and may soon appear in this country. For some time now, I've predicted that odors will be added to commercial products to enhance and change moods. Odorized products that promote both sound sleep at night and wakefulness in the morning obviously have enormous potential in the marketplace.

It's also possible that as a result of the scents, some of the men may have become alert and aware of their entire environment, including any sexual cues. This in turn may have increased penile blood flow.

Different parts of the brain control different functions and activities. From animal studies we know that if we stimulate the erectile center, or the septal nucleus, in the brain of a male squirrel monkey, an erection is induced.[5] Interestingly enough, a direct pathway connects the olfactory bulb to this part of the brain in both squirrel monkeys and humans. When an odor can act directly on the brain and cause a physiological response, this means that the odor has a neurophysiological action. Remember, one of our subjects slept through the study and still had a strong response to the pumpkin pie and lavender mixture. We suspect his response was neurophysiological, in that the odor itself, regardless of the environment or other factors, caused the increased blood flow. It's possible that the odors influenced his dream content, which in turn stimulated blood flow to the penis. Dreams cause a variety of neurophysiological changes, and scents can change the content of dreams. Because the subject had no memory of dreaming during the study, we can't eliminate dream-influenced response as a possibility.

One of the study subjects asked if we could be certain that the responses he and the other men had were directly sexually related. Well, we think they are for the most part, but we also know that some odors can induce an aggressive response, which in male squirrel monkeys (and probably human males as well) increases blood flow to the penis. Again, this would have been an involuntary response, not one that came from consciously aggressive feelings.

This still leads us to wonder why food smells, in particular, brought about the strongest response in our subjects. If we think about this, it's actually quite logical in light of our evolutionary past. The relationship between food and its odors and sex may be part of the wonderful mechanism that keeps us reproducing. When food is plentiful, sexual relations occur in a predictable pattern. We in the West don't think about this nowadays, because food is almost too plentiful—at least for the majority of people. Our worries center around eating too much, and we seldom feel a sensation of hunger for more than an hour or two. But this wasn't always true. In times of famine, sexual patterns change, too.

My sense is that early human beings spent much of their time foraging for food, and they gathered at the site of animal kills. They probably ate large quantities when food was plentiful, but their ability to preserve it was quite limited. During these times, there was opportunity and inclination to procreate. The link between food and sex was established, and as humans became, for the most part anyway, less nomadic and more tied to agricultural settlements, there remained an association between plentiful food and sexuality.

It's also possible that the very earliest humans wandered alone or in very small groups and congregated with each other when the odors of food permeated the environment. At that time, the opportunity to mate also became available, and again, the connection between plentiful food and procreation was established.

Nature has its own intricate designs. It's no accident that women require a minimal percentage of body fat in order to conceive. In times of famine, menstrual cycles may stop for a time, and infertility results. Generally, when conditions become favorable to support new life, body fat is restored, and the fertility cycles return. The connection between food odors and the brain is just one of the sexual inducements that returns in full force. In a sense, we have a procreative imperative when the time is favorable and an equally strong imperative not to reproduce when conditions are unfavorable. In evolutionary terms, plentiful food odors equal a favorable time to procreate.

These days many people logically ask, "But we are overpopulated now, and we use a dozen or more different methods *not* to reproduce, so aren't these mechanisms obsolete? And, if we're rational human beings, how could an odor affect us in such a way that we think about procreation?"

Well, I didn't say that these odors make us think about reproduction. These odors make us think about sex, and most modern people in all societies separate their sexual lives from procreation. We use every means we have to control our fertility, but except for a few isolated examples, we're not willing to forgo sexual pleasure. However, our brains are still "wired" for the reproductive imperative. The human male is still fertile for most of his adult life, and the female can reproduce until age forty or more. Our brains respond the same way they did thousands of years ago. Preservation of the species is always in question; we have built-in mechanisms to give us every chance to keep our species going. In recent years, we've done a good job—perhaps too good—of reproducing ourselves; but since we don't know what will happen in the future, our sense of smell is one of the marvelous tools we have to ensure the survival of our species. When we look at the ways of nature, individual survival is not as important as ensuring that the species continues.

If you've read my previous book, you know that food odors are an aid in weight loss for both men and women, and some food odors can change mood state. So, it really isn't so surprising that food odors affect our sexual-arousal mechanism. When we laugh at the idea that pumpkin pie and cinnamon rolls and strawberries can be part of this arousal system, we are really laughing at ourselves. We are reminded that the most sophisticated among us are being led around by the nose.

Sexual Dysfunction and Odors

Increased knowledge about our senses may have implications beyond just the fun we can have with our sexual partners. We may be able to take what we know about the sexual response to odors into the clinical setting.

The area in which we can most effectively measure results of odors on sexual dysfunction involves male impotence that is *organic* in origin. By that I mean that the cause is physiological rather than psychological. As I said previously, an erection depends on blood flow to the penis; this organ essentially becomes engorged with blood during the arousal stage. Certain conditions and diseases interfere with blood flow, and hence, the ability to achieve a complete erection is compromised.

Diabetes is the most common medical condition that interferes with sexual performance and enjoyment because the disease causes circulation difficulties. At the present time, we're experimenting with odors, such as those we tested in our previous study, to see if scents can be an effective aid in enhancing blood flow to the penis in those afflicted with diabetic impotence. This would represent a method to treat a common problem safely in a noninvasive way, that is, without drugs and their nearly inevitable side effects or surgery, such as penile implants. Given what we know about odors thus far, this notion is not far-fetched.

Could Odors Help Keep Sex Alive As We Age?

About half of all sixty-five-year-olds have experienced some loss of the ability to smell; three-quarters of the population that is age eighty and over have significant diminishment in their ability to detect and identify odors. What this means is that odors used in the environment of the elderly should be more concentrated—stronger. The same concentration of an odor that may be overpowering to a thirty-five-year-old may be pleasantly subtle to a person twice that age.

Odors are a safe treatment for older men who may find that the spirit and the mind may be willing but the body doesn't respond, at least not as quickly as it once did. If the odor of pumpkin pie slices and lavender or strawberries increases blood flow to the penis, many men may incorporate specific odors—based on individual trial and error— into their lovemaking. In our study, the very common odor of vanilla had a positive response in older men. Certainly, odors aren't magic or the cure-all for everything, but they may help us enjoy the delights of our sexuality even as we age.

I believe odors will be used in our sexual lives even more now that our population is rapidly aging. As a generation, the baby boomers brought discussions about sex into the open. (Hardly a day goes by without one of the TV or radio talk shows featuring a sexual topic.) This group of many millions of Americans is, for lack of a better term, a big market segment. Commercial interests have been catering to baby boomers since the late 1940s and will continue to do so. If odors enhance the sexual experience, then you will eventually find increasing numbers of odorized products designed to enhance romance and sexuality. Certainly, any common odor you find in sachets, candles, massage oils, or sold as essential oils in many specialty shops and health food stores are safe for most people to use.

The top-performing odors in our study will no doubt make their way to the marketplace in various commercial forms. Lavender already is available, of course, and the spices used to make pumpkin pie, such as cinnamon, nutmeg, and ginger, are available in potpourri, candles, and room scents. And nothing is stopping us from taking a package of licorice to bed. I must admit, I find it amusing to think about condoms odorized with pumpkin pie spices, but that's the way of the world. So, while the study has a light tone to it, we actually undertook it for serious reasons, and the results have serious applications.

But What about Psychological Causes for Sexual Problems?

To date, no one has found a definitive treatment for impotence, regardless of the cause. Some odors may help in certain cases, but we wouldn't claim we've found a cure for anything. If the psychological causes of male impotence are related to serious internal conflicts, then I don't believe odors will be of great help. Psychological treatment and counseling are in order, and possibly a sex therapist as well. I do predict that sex therapists will incorporate odors into the sensory-related exercises they recommend to their clients.

On the other hand, if diminished desire or episodic impotence (meaning that the problem occurs now and then and isn't a chronic condition) are related to stress and fatigue, then perhaps odors could play a role in breaking that pattern. Diminished sexual desire is not the same thing as actual impotence, but it sometimes seems like it to men who want to feel that "ol' feeling" again. There's certainly no harm in trying odors to aid in the arousal stage of lovemaking. In chapter 11, I'll discuss some scents that alter mood and may help relieve stress, which without question is part of lack of sexual interest.

When you think about what this study has to do with your bedroom, consider that what we know about odors and sex is quite significant. If lavender scent or a licorice stick can lower anxiety, help you relax, and help create a loving mood for you and your partner, then by all means incorporate these fragrances into your sex life and have a good time experimenting.

After testing our group of male volunteers, we posed questions about female sexual response. Would food odors have the same effect? Or would women respond less to food smells and perhaps more to a different category of odors? Only a research project would provide the answers. In the next chapter, we'll explore the results of our study of odors and female sexual arousal.

7

Do Women Find Any Odors Irresistible?

MORE THAN thirty years ago, the famous sexuality researchers William Masters and Virginia Johnson suggested that odors influenced sexual arousal, but at that time they had no accurate way to test this theory. Of course, other researchers focused attention on odors and their connection with sexuality, too. As previously mentioned, Sigmund Freud was concerned that if the sense of smell was not repressed, men—but not necessarily women—would walk around sexually excited all the time. Freud also linked odor with the Oedipal conflict when he said that during a boy's development, he learns to recognize the odor of both parents, and eventually he comes to dislike the father's odors and have positive associations with the mother's odors.

When viewed in the context of the developmental conflict Freud labeled the Oedipus Complex, an otherwise neutral or inert odor was endowed with a sexual connotation or a sexual energy. Although Freud also spoke of the Electra Complex, a corresponding developmental conflict in girls, he did not mention the significance of the parents' odors in female psychological maturation. In any case, Freud's work follows centuries of cultural practices that linked olfaction and sexuality. Attempts to suppress or deny the link have failed and always will. We are now beginning to understand that the relationship between odors and sexuality has clinical as well as social or cultural implications.

The Interest Isn't Frivolous

Research about male or female sexual response and the factors that inhibit or enhance it often amuses people—it certainly amuses the media. To be sure, commercial scent producers are interested, too, because perfumery is a multibillion-dollar industry. While I understand the entertainment and even economic value of the results of sexual-arousal studies, I also recognize that they have important clinical implications.

In the early 1950s, sex researcher Alfred Kinsey found in one of his surveys that 10 percent of married women never experienced coital orgasm.[1] In 1956, a British study of 3,705 women reported that 10 percent of women rarely experienced orgasm and another 5 percent never experienced orgasm during intercourse.[2] In the 1970s, a study in the United States reported 17 percent of women seen at a gynecologic clinic stated that they had difficulty achieving orgasm with a partner, and 6 percent had never experienced orgasm with a partner.[3] As previously discussed, sexual dissatisfaction is common among married couples in our society, and arousal disorders are a very common form of sexual dysfunction. While we know that reasons for dissatisfaction are many, if we can find an odor or a group of smells that enhances female sexual arousal, then we can treat a condition that has received little attention from the medical community. Male impotence is traditionally considered a more important problem in which to invest research dollars. I hope that in the future we will have the ability to develop odors that can be used as safe, noninvasive treatments for male impotence and female arousal disorders. Certainly both of our foundation's studies add information on which to build.

The Birth of Our Female Arousal Study

The results of the male sexual-arousal study discussed in the previous chapter were revealing in that we learned that odors do influence penile blood flow, and it logically follows that odors would influence female arousal as well. It makes evolutionary sense for both sexes to be affected by similar arousal stimuli. So, following the promising results of our male study, we began to design a similar study, which took place in 1997, using female study subjects.

Our thirty volunteers were recruited during an appearance on the evening news on WLS, a television station in Chicago. The response was immediate, and our telephone lines at the Smell and Taste Research Foundation were flooded with over three hundred calls. Our study design called for women between ages eighteen and forty. They needed to be ovulatory and could not be pregnant at the time of testing, nor could they be actively trying to become pregnant, although we attempted to perform the tests on individuals around the time of ovulation. The reason we established the upper age limit at forty was because ovulation is more predictably regular among women under age forty.

Anorgasmic women—that is, women who do not experience orgasm—were also excluded because we wanted to test a group of women who fell into the large normal range of sexual response. If we established the effect of odors on orgasmic women, researchers would later be able to isolate certain odors and test them as a treatment modality for women with orgasmic or arousal disorders.

Because so many medications influence sexual arousal, we excluded women who were taking prescription or nonprescription drugs, including oral contraceptives. While it's well known that drugs used to treat hypertension may have an adverse effect on sexual desire and function, particularly among men, it is less well known that antihistamines, many antibiotics, some drugs designed to lower cholesterol, and many medications with neurologic effects also affect arousal.[4]

In addition, a variety of diseases are known to affect the autonomic nervous system. The autonomic nervous system has two divisions, sympathetic and parasympathetic, both of which control unconscious functions, such as blood pressure, heart rate, and glandular and organ functions. The parasympathetic nervous system is important for both penile and clitoral arousal, and the sympathetic system is important in regulating orgasm. Therefore, we excluded from our study women who had diseases affecting either system. These diseases include: thyroid conditions, hypertension, heart disease, kidney disease, and diabetes mellitus. We also screened out women with a family history of chemosensory disorders—smell and taste disorders. Since any strong odor can trigger an asthma attack, we excluded women with a history of asthma.

We also excluded smokers, because habitual smoking impairs olfactory ability; in addition, we eliminated women who ingested more than one alcoholic drink per day because olfactory acuity decreases as alcohol consumption increases. And we excluded those using illegal drugs such as cocaine. (In retrospect, we should have had these criteria for the male study as well.)

As you can see, our criteria for participation limited the variables we had to consider when reviewing results. This was important because we wanted to establish a baseline, and results of other studies could be compared with a group who had no known medical condition or personal habit that affects their olfactory ability.

Measuring Vaginal Blood Flow

The women in our study were instructed not to have sexual stimulation, either by a partner or through masturbation, for forty-eight hours prior to testing. (We attempted to test women within a day or two of expected ovulation, but women who wanted to become pregnant were not good candidates, because we asked them to refrain from sexual contact at the very time chances of becoming pregnant were high.) Participants were also instructed not to wear cologne or cosmetics.

During the testing process itself, each participant was comfortably positioned on an examining table, and a sterile monitoring gauge, a photophlethysmograph, similar in shape to a tampon, was placed in the vagina. The monitoring device measured pulse pressure, which indicates any change in blood flow to the vagina. Other studies have used this method, and it is considered a reliable way to measure female sexual arousal.[5] The gauge was hooked up to a computer, and changes in pulse strength were recorded on a continuous basis.

A surgical mask, untreated with any added odor, was placed over the nose and mouth, and we recorded any response that the presence of the mask induced. In a double-blind, randomized manner, masks that were impregnated with different individual and combined odors were presented to each woman. Neither the examiner nor the study subject knew what odor was introduced at any given time. The odorized mask was in place for one minute, followed by a three-minute "washout" period, during which blood flow was measured when no mask was in place. The idea was that once the effect of the odor—positive or negative—was eliminated, presumably blood flow returned to baseline.

As in the male study, a variety of odors were tested, including: charcoal barbecue smoke, mesquite barbecue smoke, cucumber, cherry, lemon, banana-nut bread, pumpkin pie, lavender, Good & Plenty™ or Liquorice Allsorts candy, cranberry, baby powder, sweet pea, parsley, coconut, green apple, baked cinnamon bun, peach, Oriental spice fragrance, grape, chocolate, root beer, cappuccino, gardenia, and some popular perfumes and colognes.

After the one-minute measurement was taken, the women were asked if the odor was familiar, if could they identify it, and if they liked it or disliked it. Each woman was also given a series of tests of olfactory ability. One is the standard forty-question "scratch'n' sniff" test we mentioned in the previous chapter, in which the person is given four choices for each question. For example, in one question, an odor is presented, and the individual is asked if the odor is pizza, motor oil, peanuts, or lilac.

All the subjects also were given an odor threshold test, during which bottles containing various concentrations of pyridine, a chemical whose odor resembles scallops, were presented to the subjects. This test evaluates the concentration of the odor that must be present before it is detectable. This group of tests allowed us to establish olfactory acuity and identify those who had a normal sense of smell and those with lesser—or no—ability to smell.

Sexual Arousal/Sexual Anxiety

Our study also attempted to gather as much demographic and personal data as possible, and to that end participants completed an extensive questionnaire that asked for information about favorite colognes (and if the subject wore fragrances on a regular basis), favorite food, least favorite food, number of sexual partners and encounters in the previous thirty days, sexual preference, and so forth. We also asked if a particular odor caused the women to recall their childhood.

In addition, we asked about orgasmic functioning, including frequency in the last thirty days and over the previous six months. Through these questions we were able to determine the number of women who experienced orgasm but did so infrequently, and we also were able to identify a group who were multiorgasmic. When we started the study, we didn't know if this information would make a statistical difference when measuring response to odors, but we wanted it to be available for comparison. As it turned out, these data helped us analyze complex results.

The final group of questions was very personal and explored the study subjects' experiences with and attitudes toward various types of sexual activity. Each activity was rated for its ability to arouse or to inhibit arousal. Using a scale from -1 to 5, participants were asked if particular activities were either very arousing (5) or adversely affected their arousal (-1). Some of the activities women were asked to rate included: seeing a lover nude, a lover manually fondling the breasts, kissing of the breasts and nipples, a lover stimulating the genitals with the mouth and tongue, a lover stimulating the genitals manually, reading a pornographic story, seeing a strip show, undressing a lover, being undressed by a lover, lying in bed with a lover, passionate kissing, making love in an unusual place, the partner's orgasm, hearing sounds of pleasure during sex, seeing a pornographic movie, and so on.

The twenty-eight questions used to assess sexual behaviors were used to rate sexual anxiety. In other words, the participants were asked to rate not only what aroused them (or turned them off) but also what activities would induce feelings of anxiety—defined as extreme uneasiness or distress. The scale was reversed, meaning that -1 indicated that the activity was relaxing or calming and 5 indicated that the activity was extremely anxiety-producing. For example, circling the number 2 in response to a question meant that the activity sometimes caused anxiety or was slightly anxiety-producing. The range allowed for differences not only among individual women, but for each woman, depending on the setting. Obviously, undressing a new partner might provoke slight—or even extreme—anxiety, but when one is with a familiar partner, the same activity might be very arousing. On the other hand, some women found certain activities anxiety-producing and not arousing at all.

Similar to our male study, almost all our participants were heterosexual; very few reported same-sex partners. We also found a correlation between the overall enjoyment of sex and frequency of orgasm (almost 90 percent of the time) with frequent sexual activity, from several times a week to several times a day. These same women reported experiencing multiple orgasms several times a week.

You may notice that we collected more data from women than from men. This was not meant to be discriminatory in any way. Female sexual arousal, as measured by blood flow, has not been as extensively studied as male arousal. We attempted to gather as much information as possible about an area of female sexuality that needs more research. In addition, for many reasons, female blood flow studies are more costly to carry out, and in just one study, we were able to collect information that covered a number of different areas.

Response to Various Odors Was Complex

If you recall from the last chapter, no odor actually decreased penile blood flow. The corollary was not true with our female study subjects, however. In fact, we found that as a group, several smells impaired arousal. Those odors that had the greatest negative effect, meaning that the baseline blood flow measurement actually decreased, were cherry (18-percent reduction) and charcoal barbecue smoke (14-percent reduction). It's safe to say that we won't be seeing a rush to develop colognes odorized with barbecue smoke or cherries.

Other odors had minimal effect. For example, male colognes actually decreased vaginal blood flow by 1 percent, and female perfumes increased it by only 1 percent. A combination of baby powder and that lover's food, chocolate, resulted in a 4-percent increase.

The leaders in the male study—pumpkin pie and lavender—increased vaginal blood flow by 11 percent (as compared to a 40-percent increase in penile blood flow). The odor that had the greatest power to induce female sexual arousal was a combination of Good & Plenty or Liquorice Allsorts and cucumber, which increased blood flow by 13 percent.

As far as odors and sexual arousal go, men and women appear to be different. Results of the male study showed us that men are a fairly homogeneous group when it comes to response to odors, in that the most arousing odors were the same for the majority of men.

While the Good & Plenty or Liquorice Allsorts and cucumber combination was arousing to most women, marked differences occurred among the participants, based on the kinds of sexual behavior and activities preferred. For example, the women could be subgrouped into those who were extremely sexually aroused when a lover manually stimulated her genitals and those who were not. Women who found manual genital stimulation arousing showed a 12-percent increase in vaginal blood flow in response to pumpkin pie and lavender and averaged an 18-percent increase with the Good & Plenty or Liquorice Allsorts and cucumber combination.

Alternatively, no odors induced sexual arousal in the women who were not extremely aroused by manual genital stimulation, whereas many odors inhibited arousal, including male colognes and perfumes, both of which decreased blood flow by 14 percent. Even Good & Plenty or Liquorice Allsorts and cucumber decreased blood flow by 13 percent in that group.

We also saw differences between women who found masturbation arousing and those who did not. Among women who reported being extremely aroused by masturbation, every odor had an arousing effect. A Good & Plenty or Liquorice Allsorts and banana–nut bread combination (28-percent increase) and the Good & Plenty or Liquorice Allsorts and cucumber combination (22 percent) showed the greatest effect. Popular perfumes showed an 18-percent increase in vaginal blood flow, as did baby powder, which was nearly as arousing at a 16-percent increase.

Women who did not find masturbation extremely arousing showed an increase in vaginal blood flow of 16 percent in response to baby powder and a 10-percent response to lavender and pumpkin pie.

We also discovered differences in response to odors among women who reported being multiorgasmic during at least one-third of their sexual encounters versus those who experienced a single orgasm during their sexual encounters or reported being multiorgasmic less than one-third of the time. Among those who were frequently multi-orgasmic, the odor of baby powder *reduced* vaginal blood flow by 8 percent; the mono-orgasmic were aroused in response to baby powder with an average *increase* of vaginal blood flow of 15 percent.

Why Can't Women Be More Like Men?

Clearly, while women's responses to odors were not as homogeneous as men's responses, the Good & Plenty or Liquorice Allsorts and cucumber combination was the most effective odor overall. However, subgroups of women responded differently, depending on their preferences of sexual activities and behaviors.

We started the study with two hypotheses, both based on the effect of food odors. As described in the male study, we suggested that sexual arousal in the presence of food odors makes evolutionary sense. It is likely that our early ancestors were relatively solitary, but they gathered around food kills; hence, the potential for finding a mate was improved, and chances for successful procreation enhanced, in the presence of food. It is likely that a selective advantage would have arisen from a situation in which sexual arousal occurred in the presence of food. In addition, it does not make good evolutionary sense for one sex of the species to be aroused when the other is not. Since we found that men are aroused by food odors, it was probable that food odors would be arousing to both sexes. Our study showed that the most effective odor for female arousal was a combination of food odors, so we concluded that this hypothesis was correct.

We also proposed a counter-hypothesis. In our society, women are still largely responsible for preparing food for their families and in the workplace; hence, food odors may be imbued with the sometimes unpleasant associations of responsibility and work. We speculated that food odors might be a sexual turn-off for women with such negative associations. Indeed, some food odors did impair sexual arousal or had a neutral effect; however, the arousing action of the combination of Good & Plenty or Liquorice Allsorts and cucumber had to be so strong that it was able to overcome any turn-off effect. Therefore, our counter-hypothesis turned out not to be true.

How Do Odors Influence Sexual Response?

When we see particular responses to odors, we often look for the Pavlovian response—the conditioned response to a familiar odor. If that were the case, then the results suggest that baby powder, cucumber, banana-nut bread, and Good & Plenty or Liquorice Allsorts reminded the women of their partners! This doesn't seem likely, so we doubt that a Pavlovian response is a likely explanation for the results in most women.

It's possible that the odors evoked a nostalgic response. For example, some women reported that baby powder made them recall their childhoods. Nostalgic responses tend to induce a more relaxed mood and remove inhibitions, thereby allowing for greater sexual responses to external stimuli.

Alternatively, some odors may have acted on the brain in a more physiological way. For example, the odors may have stimulated the reticular activating system of the brain, which makes one awake and alert. When your alarm clock buzzes in the morning, it is this system that is activated. In this alert state, the women may have become more aware of sensory stimuli in the environment, including sexual cues.

In addition, the odors may have acted directly on the brain to reduce anxiety, a connection made in other research. For example, in one of our foundation's studies, we placed people in a coffinlike tube, which induced a claustrophobic response.[6] We then added odors to the environment and evaluated their effect. The odor of cucumber reduced anxiety and altered participants' perception of space. In a similar way, the odor of cucumber may have reduced anxiety among women in our sexual-arousal study.

We have two theories about the response found among women for whom masturbation is extremely arousing. As previously explained, a monitoring device was in place in the vagina throughout the test. Perhaps the odors acted to change the women's focus of attention and enhanced the perception of touch. We know that our senses act on each other. If you close your eyes and press your fingers against your eyelids, you perceive color, even though color isn't there. You are perceiving the sensation of touch as vision. Similarly, the presence or absence of one sensory modality can affect our perception of another. Many people say that they have improved perception of auditory stimuli when they are in total darkness. If you are attempting to discriminate between softly spoken words, you may—often unconsciously—close your eyes to improve your perception. In our study, the sense of smell may have acted on the tactile sensations produced by the vaginal monitor. Furthermore, since we already know that these women were easily sexually excited by touch, it's possible that the odors had an even greater effect in combination with the tactile sensation produced by the monitoring device; their touch receptors were already conditioned to be more sensitive to tactile stimuli, and the olfactory stimulation further enhanced the perception of touch.

Alternatively, rather than the odors increasing awareness of the vaginal monitoring device, the odors might have made the women less aware of the device. For some women, the monitor may have been slightly painful and the odors may have distracted them and, therefore, decreased their discomfort. Or, the odors may have acted physiologically to reduce pain. In a study we made of individuals who suffered from migraine headaches, we showed that the odor of green apple relieved pain; since pain inhibits sexual arousal, the odors may have acted to reduce the discomfort of the vaginal monitoring device.[7] When there is minimal pain and discomfort, sexual arousal is enhanced.

Finally, one explanation for both the positive and negative responses that makes anatomic sense is that the odors acted on the septal nucleus, which as already mentioned, is the erection center of the brain. Animal studies have shown that stimulation of the septal nucleus of the squirrel monkey results in erection, and a direct anatomic connection exists between the olfactory bulb and the septal nucleus. The odors may have acted directly on the septal nucleus to either stimulate or inhibit arousal.

Where Do We Go from Here?

The results of our male arousal study showed that a wide variety of odors have a positive effect on penile blood flow; the results of our female study are more specific, and what we've learned is that women's responses to odors are more complex and vary based on preferences. We'll need more research to pinpoint an odor that evokes a universal response in women, if indeed such an odor exists.

For now, we know that men are wise not to make assumptions about the effects of their favorite sexual smells on their partner. Generalizations about odors and female arousal are not necessarily true for individual women. It appears that a man needs to learn what specific behaviors are arousing to the woman he is with. Is she or is she not sexually excited by manual stimulation? Does she find masturbation arousing? Is she generally mono-orgasmic, or does she have multiple orgasms much of the time?

Remember that, overall, men's colognes do not appear to have a universally positive effect, and for some women, they actually inhibit arousal. If a woman says she doesn't like your cologne, believe her! If you don't listen, you may be interfering with her pleasure during a sexual encounter with you. And, the next time you decide to surprise your lover with a gift of candy, skip the chocolate creams and pick up a box of Good & Plenty or Liquorice Allsorts!

Our two studies established a link between food odors and sexual arousal, and since humans can detect about ten thousand odors, it is highly probable that many other odors contribute to sexual arousal. The Smell and Taste Research Foundation will continue to study this important area of human life.

In the next chapter, we'll look at the way food has—through the ages—added "spice" to our bedrooms.

8

Spice Up the Bedroom with Sexy Food

TWO OF LIFE'S greatest pleasures are eating and making love, so it shouldn't surprise us that these activities are, at least on occasion, combined. Even when not directly linked, we tend to associate food with romantic feelings, and often a young man in love brings his lover chocolates—or flowers or wine. Even in this age of equality between the sexes, an expectation remains among men and women that it's still *his* role to bring the chocolate candy and the flowers to *her* as part of the courtship ritual.

The most popular date is still the dinner date, and when a couple want to spend some time alone, or they plan a celebration for a special occasion, they have what we call a "romantic dinner for two." Sharing food also plays a role in creating an accepting environment. One reason many business transactions take place over dinner is that food facilitates what is considered persuasive conversation. When talking over dinner, couples may find that the food itself helps them discuss issues they may not feel inclined to bring up at other times.

Many couples with small children around must make special arrangements to be alone, and most find it's worth the trouble and expense to do so. Couples discover that their time spent lingering over dinner is a chance to reestablish their relationship and have some adult conversation. (Even if this time alone isn't a prelude to a sexual evening, I urge you to make sacrifices if you have to in order to get this time for yourselves. Your relationship will be better for it, and it's a natural break from the stress and worry that may dominate so much of your time.)

Dinner dates and romantic gifts are symbols that are meant to hint at the possibility of intimacy on many levels, sexual intimacy included. It's no accident that many romantic gifts are aromatic, but if the taste of the food we're eating during an intimate dinner is to be fully enjoyed, we must be able to smell it.

Odor and Taste Can't Be Separated

It often surprises people that 90 percent of "taste" or "flavor" is actually smell and that the odors in our food provide our sensation of its taste. You can test this for yourself. If you hold your nose and take a bite of chocolate, the sweet, sensuous candy suddenly tastes like cardboard. Those whose ability to smell is diminished usually spice their food heavily in order to get enjoyment from it. Women, whose sense of smell is better than men's, generally enjoy food that is less heavily spiced than the food men prefer. When women wonder why some men reach for the salt and pepper shakers before they even taste the food, it's often because they have become accustomed to being served food that is usually, by their standards, too bland. Hence, they must add true tastes—salty, sweet, bitter, and sour—to compensate for their poor sense of smell.

Most of us take our ability to taste and smell for granted. We can distinguish between the taste of a juicy, sweet apple and a more bland potato, and we enjoy the spicy or sweet dressing on our salad. But take away the ability to smell, and an apple and a potato are barely distinguishable, and a salad dressing is just bland oil.

About two million people in the United States suffer from a diminished ability to smell. I call this a hidden disability, because few people talk about it, and fewer still have much sincere sympathy for those who suffer—and believe me, people who can't smell do suffer. As I've said, at our foundation we typically see patients who have come to us because they have olfactory loss as a result of a head injury sustained in an auto accident, but there are many diverse causes of diminished ability to smell. Nutritional deficiencies, side effects of medications, and even AIDS can reduce olfactory ability.

As I've pointed out, a normal and gradual loss of olfactory acuity occurs as we age. By age sixty-five, about 50 percent of the population is affected, and the numbers rise to 75 percent by age eighty. This gradual olfactory diminishment may have an effect on sexual desire, as well as on our ability to enjoy food. Some elderly individuals complain that they have lost their appetite and no longer derive much enjoyment from their food. This can have serious health consequences, and yet this situation usually isn't addressed adequately by those specializing in the care and treatment of the elderly. We know that quality of life diminishes as the ability to smell the odors in the environment decreases.

I've digressed a bit here, because the ability to enjoy life to the fullest is dependent on our ability to smell and, therefore, taste. These two senses are clearly linked to our enjoyment of sexuality. One of the possible consequences of the loss of the ability to smell is loss of interest in sex, regardless of age.[1] It's as if the pheromones, or sexually stimulating odors in the air, aren't detected, and the special scents associated with one's lover and with sexuality in its totality are no longer present. The chemical "signals" literally floating around in the air can't work their magic. Gradually, the person realizes that he or she is not thinking about sex as often, and it's difficult to enjoy the symbols—including scents and tastes—of romance. Eventually, one forgets about the rich fragrance of roses or the smell and taste of chocolate or the aromatic bouquet of wine.

I mention these three particular symbols—flowers, chocolate, and wine—because they are so common in our culture. Perhaps other items have deeper personal meaning for you, so substitute your own special symbols. In any case, I would venture to say that whatever symbols you associate with romance and sex probably have tastes and smells involved. (If you have found your desire for sex waning, consider that your sense of smell is no longer as sharp as it once was. If you've lost your sense of smell completely or it has diminished significantly, you may want to be tested, because you could be one of the fortunate individuals whose loss of ability to smell can be treated and restored.)

Adam and Eve and the Apple

In the mythologies that have shaped our culture, fruit has often been linked with and has symbolized female sexuality. Our famous creation story is no exception. Eve tempts Adam with fruit, a nearly universal symbol of female sexuality. The serpent, or snake, helps her lure Adam, who succumbs, and we all know the rest of the story. (In Western culture, influenced by Freud, the snake is a phallic symbol; in other cultures, the snake is a symbol of feminine wisdom. Since both work in the myth, pick your favorite interpretation.)

In some folktales, the flowering apple tree represents femininity and fertility; in other tales, eating a ripe, soft pear is symbolic of partaking of feminine wisdom and creativity. The pear and other fruit, such as peaches or apricots, are also viewed as representative of female genitalia. Figs, on the other hand, are said to be symbolic of male virility; if we look at their shape, we can see why.

Both flowers and fruit have been associated with female genitalia. In fact, one of the common expressions for first-time sexual intercourse for women is "deflowering." In some marriage ceremonies within the Catholic tradition, women offer flowers to the Virgin Mary and are symbolically offering their virginity to the person they revere as the mother of Jesus. This ritual is by no means as common as it used to be, but it speaks loudly to the sexual association of the "flowering" of a woman's sexuality and her ability to bring forth the "fruit" of her womb. Consider, too, that a woman is said to be barren if she's infertile, much the way land is barren when it doesn't bring forth its fruits.

In dreams, fruit is often associated with sexual images and fertility and may or may not have a direct connection with the dreamer's sexual expression. For example, images of fruit and opening flowers may represent creativity, which of course, is linked with the feminine in folklore and in many cultures.

Some dream analysts view the image of an opening flower in a dream as a universal symbol for sexuality. In the *Kama Sutra*, the ancient Indian book on the art of love, one of the positions for sexual intercourse is called *Uptahalla*, translated as the "Opening Flower." (We in the West know this same position by the far less poetic "missionary position.") Another is called *Ratisundara*, or "Sweet Love."

The Chinese concept of yin is associated with the feminine principle and right-brain activities, and the yang concept is associated with the masculine and linked with left-brain functions. In general, fruit is considered a yin food and is linked with the expansiveness and creative power of the female principle. Dense, hard foods, such as root vegetables, are considered yang foods and are more representative of the male principle. (The philosophy of yin and yang is far more complex than this, but for our purposes, this simple illustration shows the universal nature of some images and associations.)

A year or two ago, research results were released about the risk of heart disease, and scientists described the varying risk factors between "pear-shaped" women and "apple-shaped" women. This amused me because our historical and cultural past and its symbols again made their way into the world of "facts" and science. Any number of terms could have been used, but somehow, pears and apples emerged.

Feasting on Each Other

Apples and pears aside, our language is filled with sensuous images of food and sex. In his popular sex manual *The New Joy of Sex*, Dr. Alex Comfort divides the sections of the book in a recipe or menu fashion, starting with "The Ingredients" and moving on to "Appetizers," "Main Courses," and "Sauces." Food is included in the "Appetizers," implying that a meal is the traditional and still very common prelude to sex.

We also hear people say, "He's so handsome I could eat him up," or "I could just devour her." We talk about our sexual "appetites," and some people might say they're "hungry" for love. I could say that I wrote this book to help you "spice" up your sex life. Or, you could tell a friend you're in the midst of a "juicy" love affair.

We use our mouths to eat and also to show affection and make love; the connection couldn't be more clear, given the many words and phrases we use to link food and sex. Pheromones and other scents may attract us, but as we begin to make love, we not only inhale these odors, we begin tasting each other, too.

Food sometimes is brought into our bedrooms because we want to enhance the sensory delights of lovemaking. We may feed each other because love and romance arouse the need to nurture and care for each other; so we feed each other as we feed children. But adult "children" use this action as a way of showing affection, often playful affection that leads to passion. Unfortunately, as couples become accustomed to each other and their romance cools, they seldom devote as much time to these pleasurable acts that enhance sex as they did when their love was heating up.

An otherwise happily married woman whose sex life was, as she put it, "poor," sounded wistful about the sexual adventures she had once enjoyed. She contrasted the fun of the past with what her sexual relationship with her husband was currently like. "We used to do so many romantic things—we'd make love in our small garden behind the house or we'd take a picnic to the beach and feed each other finger food and then we'd walk back home and make love. We used to take time for quiet dinners, and sometimes we'd take champagne and dessert to bed. I can't remember the last time we did any of that. When I mention a picnic, my husband mentions sand in the chicken and ants in the salad."

If this sounds familiar, even if the details differ, then you're not alone. Certainly, some of the things we do when we're newlyweds or experiencing the freshness of a new relationship contribute to a lasting bond between two people. The problem is that we do let romance die or we save these pleasurable activities for such special occasions that our attempts at romance run the risk of seeming contrived and mechanical.

I've heard many women complain that men gradually lose interest in these romantic things, and they feel neglected. It's as if the "conquest" has been made, so he doesn't need to wine and dine her anymore. Some women complain that their husbands even watch television during their lovemaking! Unfortunately, women who feel emotionally neglected eventually begin to resent clumsy and quick sexual advances, and they miss the romantic attention. Men are often perplexed and may protest that they don't love their wives any less, but it's just that they're preoccupied with other things.

In truth, men often miss the romance, too, but they may be less willing to admit it. One obvious and very easy solution to this dilemma (in addition to turning off the television) is to bring the tastes and smells back to your sex life, with an attitude of playfulness and fun. For example, after our conversation, the woman mentioned above brought a bottle of champagne and two glasses into the bedroom and began drinking it herself. Her husband was puzzled, and he asked her to pass him a glass. "It's been a long time since we've done this," he said, and he began to reminisce about their courtship and newlywed days. Soon they were laughing and talking over old times.

I could say that this evening ended in blissful sex, but that's not exactly true. Neither partner was accustomed to drinking champagne, and to put it bluntly, the alcohol went to their heads, and they fell asleep curled up together. But, the next day they had blissful sex— she brought a bowl of fruit to the bedroom, rubbed a strawberry on his back, and proceeded to lick off the juice.

This woman's story illustrates some important points. Consider the following list of tips for using food and their odors in the bedroom.

Use alcohol in moderation or not at all

In *small amounts*, wine and champagne are sensuous and add to romance. Too much of a good thing, though, and romance becomes hazy and unfocused. As men grow older, they often find that more than one or maybe two alcoholic drinks interferes with achieving erection. The sexual exploits of heavy drinkers we see on the movie screens are illusions. "Reel" people aren't like the rest of us mere mortals. Shakespeare best summed it up when he warned: "It [alcohol] provoketh desire, but takes away performance."

We also know that alcohol is a central nervous system depressant and slows reaction time and, particularly when consumed to excess, diminishes judgment as well as pleasure. Remember that it also reduces olfactory acuity, which decreases sexual performance and even desire. Because of its desensitizing effects, Dr. Alex Comfort advises refraining from alcohol completely during an evening when sex is on, as he might put it, the "menu." (That's a sexier word than "agenda.")

While I'm not suggesting that you must completely abstain from alcohol, I will say this: If you're planning a romantic dinner or are taking wine to bed, enjoy the aroma, take time to inhale the fragrance, and take small sips from the glass. That's what wine connoisseurs do. In a survey of Americans by the Revlon Company, wine was listed as a sexy food by over one-third of adults.[2] (Beer and hard liquor weren't considered sexy, however.) This leads me to believe that unless there is a medical reason (such as alcoholism, obviously, or diabetes) to abstain, a small amount of wine or champagne is harmless and may help partners relax. In very small amounts, alcohol can lighten your mood and put the cares of the day aside and, therefore, help you to be in a more receptive mood for sex.

Some people believe that wine and champagne are particularly associated with romance and sex because of the types of glasses we usually serve them in. It is said that the long stem and deep bowls containing the liquid may subconsciously remind us of male and female genitalia. Even the way we hold the glass can be suggestive of erotic caresses, so even if you're using sparkling water as your beverage, serve it in wine or champagne glasses. If nothing else, it will make you feel like you're enjoying a special occasion.

Finger food is generally the sexiest food

Strawberries and chocolates, also mentioned by a significant number of people in the Revlon survey, are finger foods and are easy to feed to each other. Grapes, crunchy nuts, and even raw vegetables and dip can easily be eaten in bed. Of course, these finger foods are not just for the bedroom—and neither is sex.

Most sex therapists agree that one of the easiest ways to spice up an otherwise normal sex life is to vary the time and place of your encounters. If you have privacy in your living room or some other room in your home, then use food to enhance the romantic setting. Bring a dessert of fruit or chocolate—or bite-size pumpkin pie pieces or licorice sticks, of course—to the living room. Keep the television off! Pay attention to each other, and have some fun with the food.

Some lavender, in an odor diffuser or as scented candles, is a good idea, too.

Expand the idea of "edible" sex

Placing food on the body is nothing new, and many lovers play this kind of food-sex game. Strawberry juice on the back is just one of the wonderful possibilities. The favored foods for this kind of activity tend to be whipped cream and honey. Sometimes fruit is used in combination with whipped cream or honey, and the lovers feed each other and then put the food on their bodies and eat from the skin as they would from a plate—minus the knife and fork, we hope. According to Teutonic tradition, a newly married couple was supposed to drink honey wine for thirty days after their marriage. This was said to give them a boost of sexual energy, and I imagine it tasted good, too.

Some health food aficionados tell me that honey-sweetened yogurt is a good substitute for real whipped cream, which is, let's face it, all fat. (Some commercial substitutes are not high in fat, however.) It's not surprising to me that whipped cream, which is usually flavored with vanilla, is popular with lovers. Its creamy texture is appealing, and vanilla is one of the scents that increased penile blood flow in some men in our study. I've also heard a few people mention molasses as a sexual treat because of its slightly spicy taste and its rich aroma. So, take your edible fun wherever you find it.

In the foreplay stage, the food or whipped cream or honey is placed anywhere on the body, and the kissing, licking, and mock biting involved in ingesting the food are arousing to many people; the lovers can often create a playful atmosphere. As lovemaking progresses, however, play can turn to passion, and these foods are often applied to the genitals and the stimulation involved in "consuming" the whipped cream or whatever can lead to orgasm for both men and women. As you can see, the smells and taste work together with touch and sight (and sound, too, if you choose) to create a total sensual experience.

In recent years, edible body paint has become available. It combines the sensuous strokes of massage with the sensation of removing the "paint" from each other's body through licking or kissing. As we learn more about specific odors and their ability to arouse, I expect that we will see the number of flavors grow. Right now, you can satisfy your chocolate cravings and have sex, too, because chocolate-flavored body paint already is on the market.

Some lubricants are edible, too, which allows lovers to combine oral sex with massage. Almond oil is one such example; according to popular aromatherapy literature, the scent of almond oil is considered particularly arousing. In fact, King Solomon's garden is said to have had almonds and other foods that created a sensual atmosphere for his visits from the queen of Sheba. (We didn't test almond scent in our male arousal study, so I can't back up King Solomon with scientific evidence.)

I don't recommend these "food-based" oils as lubricants for intercourse, however, because they lead to the possibility of vaginal yeast infections or even bladder infections. But almond oil serves as a safe and edible massage oil.

Some condoms are touted as flavored, but frankly, most people agree they don't taste very good, so I'm not giving them a recommendation here. And one caution: oil-based lubricants are not appropriate for use with a condom because the oil will quickly destroy the latex, which lessens its contraceptive properties as well as its ability to prevent the spread of sexually transmitted diseases. Lubricants used with condoms must be water-based, and some are considered edible, although not necessarily pleasantly flavored. Astroglide is said to be slightly sweet and is considered safe for most people to taste— although I don't mean that you should purposefully ingest the product in great amounts.

An additional cautionary note: sexually transmitted diseases, including HIV, the virus that causes AIDS, can be transmitted through oral sex, so if you are using safe-sex techniques, you should not experiment with edible lubricants during unprotected oral sex.

When combining food and sex, think "lite"

Have you ever felt like a passionate romp in the bedroom after eating a Thanksgiving dinner? I didn't think so. While food and sex are linked, too much food—like too much alcohol—can dampen desire. Steaks that weigh a pound and potatoes the size of footballs are not conducive to a night of romance. Whether we're interested in sex or not, we feel better if we eat a light dinner. One reason many people enjoy sex in the morning hours is because they are not overfed and feeling "stuffed" when they awaken.

A romantic evening does not need to go from a romantic dinner straight to the bedroom anyway. If you find yourself sleepy after a meal, take a walk in your neighborhood or do something active for an hour or so. If your evening is hectic because you have children's homework to supervise or you're concerned about your teenager who has the family car, then don't pressure yourself to feel up for romance. Instead, skip dessert and take some light, but sensuous, food to the bedroom with you when the family has settled down.

Food as Aphrodisiac

Folklore originating in a variety of cultures has provided us with tales about many foods said to be aphrodisiacs, meaning that they purportedly increase sexual desire, improve performance, and release inhibitions. To date, no scientific studies have shown that any of these foods are true aphrodisiacs. However, the mind is powerful, and the placebo effect operates; hence, people believe they are more sexually powerful if they eat these foods. If one learns as a child that oysters make you sexier, then this belief may in fact make a person feel sexy after eating oysters.

The word "aphrodisiac" is derived from the name of the Greek goddess Aphrodite, who was said to be born from the sea. In cultures influenced by Greek mythology, and this includes all of Western culture and much of the Middle East, seafood was considered to have powerful sexual properties because the odors of female sexuality are slightly salty and "fishy" and are subconscious reminders of the sea and seafood. Therefore, according to the folklore of aphrodisiacs, men are attracted to such foods as oysters and link them with sexual arousal and desire.

Associating sea odors with sex is common, and one reason that commercial odorized products use references to the ocean and sea breezes in their names is that the symbolism of the beach and the ocean have wide appeal. The smells of the ocean are soothing, which is one reason the great "getaway" vacations are usually located around the beach and water. Many of the places considered the most romantic spots in the world are on the water, from the French and Italian Rivieras to the beach resorts in the Caribbean Islands and Hawaii.

Another reason oysters are considered a sexy food is that they are often eaten live and, therefore, are said to be life-giving. The same is true for caviar. Poultry eggs probably seem too common to be romantic to Westerners, but various concoctions using them are mentioned in detail in *The Perfumed Garden*, a book I've mentioned previously that discusses the art of lovemaking. (Like the *Kama Sutra, The Perfumed Garden* is considered a classic work on sexual expression and desire.)

Oysters are also said to resemble female genitalia, as are mussels and bearded clams. Asparagus and ginseng, on the other hand, are said to resemble male genitalia. Ginseng is used widely in Asia and more recently has made its way to Western markets; some varieties are used as a tonic to stimulate hormone production. Regardless of its potential value medicinally, it is true that the ginseng root resembles male genitalia.

Other plants are said to have aphrodisiac properties, including cinnamon and ginger, which interestingly enough, are two spices commonly used in pumpkin pie. Also consistent with our study, licorice water is purported to be an aphrodisiac. If you recall from chapter 4, celery, parsnips, and truffles contain androstenol, which is considered a potential pheromonal substance.

In certain cultures, very hot, spicy foods are connected to sexual passion and machismo—that is, exaggerated male sexuality. The ability to eat extremely hot chili peppers is a way to prove one's manhood and, by extrapolation, one's sexual prowess. In those cultures, eating these hot peppers is also a ritual, a rite of passage, if you will.

The cocoa plant also takes a prominent place among the purported aphrodisiacs. The reason for this is quite interesting, and we could say it's all in our heads. Our brain chemistry changes when we fall in love, which is why some people seem to be "walking on air" when love hits hard. Some popular psychologists say this change in brain chemistry makes us act a bit "crazy," although that description is not particularly scientific. Chocolate contains phenylalanine, which is an amino acid that stimulates the production of endorphins, a brain chemical that dulls pain. Apparently, people who have just "fallen in love" show elevated levels of this amino acid, indicating that their brain chemistry does change. On the other hand, low levels of endorphins are also associated with depression. People who crave chocolate are probably looking for a mood-lifter, consciously or not. Certainly, anything that lifts our mood has a positive effect on our sexual lives, and the change in brain chemistry may help temporarily distract us from other concerns.

Some aphrodisiacs mentioned in the folklore are created from various animal substances, most notably the testicles of sheep and bulls.

Other substances are extracted from the sexual organs of mammals—and other mammalian organs, too—and made into tonics and teas. As with other aphrodisiacs, what effect they have probably comes from cultural conditioning, and that leads to a powerful belief in the substance as a sexual tonic.

Aphrodisiacs Start in the Brain

Ultimately, we enhance our sexual lives by having an open, playful attitude and willingness to experiment with ways to enjoy our sensuality. We are meant to enjoy both food and sex—it's nature's way of making sure we survive and reproduce. When we mingle the two activities, we add to the pleasure of both. So, bring on the edible body paint and the champagne and have fun. You'll be glad you did.

9

The Allure of Perfume

PERFUMES, with all their variety and uses, are at least as old as recorded history. Chinese writings, believed to be eight thousand years old, mention the use of perfume, and considerable evidence exists showing that the ancient Egyptian, Sumerian, Aztec, Greek, Roman, Indian, and Hebrew cultures were extravagant in their use of fragrances in their cultural rituals.

Although trends and beliefs about using fragrances change, at no time has perfume completely disappeared from the cultural landscape. For a short period during the Puritan domination of England, the law forbade women to wear perfumes and other scented substances precisely because of the belief that odors are such powerful sensory stimulants that their use should be banned. Occasional attempts to suppress the sensuous odors of perfumes and other odorized products have always failed precisely because scents stimulate and influence mood; this is a pleasure too powerful to resist.

Never before has such a tremendous array of choices of scents been available for personal use on our bodies and clothing, as well as in our homes. Take a trip through the aisles of a typical department store, and you'll see hundreds of perfumes, colognes, powders, and other cosmetics marketed primarily for the power of their fragrances. Fashion

THE ALLURE OF PERFUME

and cosmetic trends come and go, and new scents are continually added, too. During the winter holidays, multimillion-dollar advertising campaigns dominate television and print media, and men in particular are expected to take on the challenge of choosing the perfect scent for the woman in his life. Perfumes and other odorized cosmetic products are linked to romance and sex, but they are also linked to the less complex desire to wear fragrances that simply make us feel good.

The psychology of cosmetic use of scents is complex, but ultimately, the reason we add odors is to make ourselves more pleasing—to ourselves and to others. Good smells are also associated with positive traits; traditionally, even godliness has been linked to sweet smells, which is one of the reasons that religious services often include incense or flowers. When a man gives a woman perfume, he is sending the message that she smells good, therefore she is good. The same is true when a woman gives aftershave or men's cologne to the man in her life. She is telling him that he fits the positive image of the scent.

Perfumes and colognes and aftershave lotions are often gifts that carry sexual connotations, and for this reason, they tend to be very personal gifts. Of course, a sister can give her brother a bottle of Michael Jordan's "Rare Air" or "Cool" cologne, and she's telling him that she thinks he's a terrific guy who will attract the opposite sex. In many contexts, a gift of cologne and perfume is a way to extend a compliment and affirm a woman's femininity or a man's masculinity. It's no wonder that, worldwide, colognes and perfumes and closely related products such as powder and body sprays constitute a lucrative business.

It's been reported that, over a lifetime, the average woman will wear over fifteen hundred dollars' worth of perfume. Men have not caught up yet, but give them a few years, and they will. In the first six months on the market, Michael Jordan's line of scents grossed sixty million dollars in sales. Elizabeth Taylor's perfumes have grossed millions as well.

Positive Image and a Household Name Equal Success

Jordan and Taylor are household names throughout the world. Taylor is reputed to be one of the sexiest women to grace the silver screen and is highly regarded for her role in educating the public about AIDS and raising research funds to fight the disease. Jordan is popularly viewed as simultaneously not quite a mere human being (hence, "Rare Air") and a person of solid character and charm, whose pursuit of new challenges both on and off the basketball court is legendary. It's no accident that one of Taylor's scents is named "Passion" and one of Jordan's scents is named "Fairway Grass." Most men know they will never be champions on the basketball court, but perhaps they could give Jordan a challenge on the golf course. Every woman can be passionate, both in the sexual sense and about their beliefs, even if they don't resemble Taylor.

If Jordan had named his men's cologne "Sweatsuit," his profit-and-loss statements might look quite different. If Taylor's perfume had been called "Prim," it would have failed dismally. The competition for perfumes and other scented products is so intense that millions of dollars are spent marketing the image each presents, not just the scent itself. Our sense of smell is unique in that we rely on language borrowed from the other senses to describe it. Perfume is described as woodsy, sweet, sharp, sexy, vibrant, fresh, and so on. These words create the image of scented products, and the language associated with them is a significant factor in commercial trends and preferences.

No single cologne is considered appropriate for every occasion. This is especially true for women, few of whom choose a cologne associated with romantic nights to wear to the office. Other scents marketed to women promise a lighter, more professional fragrance. But what is a "professional" scent? The very concept sounds odd. The easiest answer is that we are conditioned to associate scents with a context. Scent developers know this and name and package their fragrances accordingly. Therefore, if you and your lover are conditioned to associate a particular perfume with romance and lovemaking, then it isn't likely you will wear it when you have a business lunch—except if your business associate also happens to be your lover! Naturally, we consider the scents we apply to our bodies to be pleasant, or we wouldn't use them; but the effects they produce differ, depending on the properties inherent in the fragrance itself and our conditioned response to it.

Are Some Odors Really More Romantic than Others?

Marketing efforts and conditioned responses aside, it appears that some odors are universally linked with femininity or masculinity, freshness or earthy sensuousness, romance or passion. Cross-cultural research has shown that, for example, plant odors, particularly herbs, tend to be used when a clean, fresh odor is desired. Eucalyptus, marjoram, mint, rosemary, camphor, and lavender are considered fresh and natural. The scent of eucalyptus is typically found in bathrooms and kitchens, and the desired result is to give these rooms a scent that connotes cleanliness, but without adding an antiseptic smell.

Historically, herbs have been used as insect repellents, and many of these strong, herbal odors were also thought to drive out illness. For example, herb-based fragrances were used during the Black Plague years ago in Europe to drive out the invisible "evil" that caused the devastating disease. At one time, virtually every culture developed a belief that disease or pestilence was transmitted through odors and that harmful "demons" could be driven out by overpowering them with good odors. Using herbal scents in this way illustrates the belief that good smells reflect holiness and that those afflicted with various diseases and afflictions have been polluted by Satan, the symbol of evil. Given that belief, it's easy to see why great pains were taken to use herbs and other plants whose odors were strong and associated with freshness and "goodness."

Now, many of these same fragrances are linked with the freshness of the natural world, and therefore, they are often used in those men's scents that are packaged to connote rugged outdoor activities and adventure. For many men, an outdoorsy image is one they like to project, even if they never leave their city neighborhood. Scents formulated for men frequently feature freshness because women prefer clean smells, and men wear scents to appeal to the opposite sex. Women, on the other hand, tend to wear cologne because it makes them feel good. Just as women choose other cosmetic products and clothing because they like the items, women choose scents primarily for the pleasure it brings them, including, indirectly, pleasure from the way others respond to them.

In the science of perfumery, odors—or groups of odors—are called "notes." The dominant odors, or "themes," in a prepared scent are called the "top notes," and the middle and lower notes are added for accent or to add to the mood created. For example, in herbal-based fragrances, the dominant notes are derived from herbs and sometimes other plants—or a synthetic version of these herbs. The effect a perfume is believed to have is based on the dominant notes. Today's consumers are enamored with fresh, "outdoors" odors, which means that herbal scents are quite popular right now. The subtext, or underlying message, is that wearing this type of cologne or perfume means that you are adventurous, are up for a hike in the mountains (or at least a round of golf), and, of course, that you're clean!

The Musky Notes

Musk dominates fragrances that are marketed for sensuous, sexual qualities. This isn't accidental, because the musky odors are closely linked with what we believe to be pheromones. For example, truffles—which in cultural folklore are said to have aphrodisiac qualities—emit a heavy, musky odor and also contain the chemical androsterone, a steroid secreted in male sweat. As you may recall from chapter 3, androsterone is one of the substances used in human pheromonal research.

I can't cite scientific proof that the musky scents have aphrodisiac qualities or are even attractants, but they are marketed for their connotations of sexuality. Our conditioned response may be activated, because wearing a scent linked with sexuality can induce a sexual mood, not necessarily because of the odor itself, but because of the expectations or beliefs about the fragrance. Nothing is unusual about this response, and in fact, cross-culturally, musky odors have been used as the base of many cosmetic preparations associated with sex and seduction and the mystery of sensual life.

Natural musk is found in glandular secretions of the musk deer, the civet cat, and beavers. Ambergris is found in the intestines of the sperm whale, and it, too, has a musky, sensual smell. These natural, musky substances have been used for centuries as the base of many perfumes. When the odor of musk is too heavy, however, the fragrance can be overwhelming and evoke an impression of uncleanliness. In other words, a light musk base may be sensual, but too much of a good thing produces the opposite effect.

Flowers Are Always in Style

Because flowers are associated with femininity and fertility, floral odors have been popular for women throughout history. From bridal wreaths in India to spring festivals in Europe, floral scents are used to enhance femininity. It's interesting that many floral odors, while considered pleasing to both men and women, are not necessarily linked with sexuality. In many cultures, floral scents are appropriate for young women who want to attract suitors but not necessarily lovers.

Currently, the light, floral fragrances are often marketed as afternoon scents, perhaps even odors that can be worn when a professional image is desired, as long as the scents aren't too sweet or heavy. So far, floral odors have not found a place as top notes in men's fragrances, probably because flowers are universally associated with the feminine qualities of nature. Even the qualities of goddesses are symbolically linked with flowers. Aphrodite herself was said to be adorned with the blossoms of roses, narcissus, hyacinths, lilies, and violets.

Fruit Makes Its Appearance

In our culture, fruit odors have not been particularly popular as perfumes and colognes, although this has changed somewhat over the last decade or so. We've used fruit-scented household products, and if you browse through specialty shops and gift stores, you'll see an assortment of apple-, berry-, or peach-scented sachets and candles. The scent of coconut is now added to suntan lotions, probably because of the exotic quality of the odor. Fruitlike odors have been used to create fragrances, primarily for women, but not usually as the dominant smell. Again, this is changing, and some fragrances marketed for their sophistication feature a fruity odor as a top note.

In other cultures, however, fruit has been a popular attractant odor and is also associated with menstruation and fertility. On one South Pacific Island, men believed that smearing a red ground cherry on the body could entice women; in Egypt, bananas were thought to have aphrodisiac qualities. Looking at our evolutionary past, it's possible that fruit odors became associated with women because they were the gatherers, and the exposure to the juices and oils of berries, nuts, and other fruit and plants were transferred to their bodies. Hence, the odors of herbs, fruit, and flowers became linked with female sexuality and fertility.

In perfumery and in popular aromatherapy, citrus odors are in a category of their own. In our culture, citrus odors, particularly lemon, are added to household cleaning products and room deodorizers, so the scent of lemon is not linked with romance and sex. Other citrus odors, such as orange and lime, tend to be light but invigorating odors, and we are seeing more cosmetic products scented with them. Certainly, orange has a long reputation as a sensuous scent.

The Scent of the Earth

Earthy fragrances are known to linger and quite literally remind us of roots and soil, of mosses and other plants. These odors are often called mellow, deep, warm, and full-bodied, much the way a fine wine might be described. If a fragrance is marketed for its natural, earthy qualities, then it likely contains rich plant odors. When mixed with additional smells, such as lavender or other floral scents, these products are said to be invigorating and also symbolize adventure and a bold approach to life. Men's fragrances often feature these odors as dominant notes; women's perfumes that contain the earthy smells are usually offset with a floral or green top note.

Spiciness and Mystery

Movies and books that feature sexual tension are often called spicy and mysterious, which is the mood that the spicy fragrances are intended to evoke. Generally mixtures of both sweet and spicy, such scents are popular throughout the world. Clove, cassia, cinnamon, and bay laurel are used in many odorized products, and historically, they were often used to mask the smell of decay. This doesn't sound romantic, but spicy odors, some of which are floral or mixed with floral odors, masked the smells of decaying food or even decomposing bodies. Egyptian embalming practices relied heavily on spices, particularly cinnamon and cassia.

Oriental spices, especially ginger, are used in fragrances meant to be haunting and mysterious. A spicy fragrance is sometimes added to a light floral odor to give it more "character" or pungency. The heavy odors of spices are rarely used as the top notes of perfumes; in most cases, they are mixed with lighter odors. Still, the perfumes that fall into the category of Oriental spice are viewed as the most mys-terious fragrances.

It's interesting that most commercial baby powder is scented with fragrances from the Oriental spice family, which has not relegated it to the world of infants and childhood. In fact, it's possible that the popu-larity of such perfumes as "Tabu" and "Shalimar," which are dominated by the Oriental spice family of scents, is related to an unconscious, nostalgic response. Most people have a positive association with baby powder—and babies. We recreate that pleasant response in adulthood by wearing a scent that, while not identical, is similar.

For the most part, wearing perfume has been considered a tradi-tional "rite of passage," much as wearing makeup is for girls and shaving is for boys. Today, however, some cosmetic and fragrance companies are marketing scented deodorants and other personal care products designed especially for children. Even a chocolate-scented perfume is available! I'm not recommending that you run out and purchase scented cosmetic products for your children, but the existence of such products signals the expanding world of commercially available odors, a world that even includes perfumes formulated for pets. In years past, most people settled for the absence of unpleasant animal odors, but a growing market for pet perfume now exists.

Other Scents that May Be in Your Favorite Perfume

Some of the odors considered woodsy and earthy are traditionally found in incense. These odors are long lasting and are made up of a variety of substances, including sweet-smelling resins, such as myrrh and frankincense. Every known culture has used various rich and strong odors as offerings to the gods and in many religious rituals, and some of these odors were popular perfumes as well. Today, these heavy and resilient odors are sometimes added in very small quantities to certain perfumes, but they aren't dominant, even in the strongly scented products designed for men.

Another separate category includes smells called "green," which certainly illustrates the nearly universal problem of describing odors. As I've said, you'll notice that we use words from the other senses to label odors. Green smells are intended to remind us of freshly cut grass and other fresh scents. Some of the green odors are floral or woodsy smells, but because of other associations and uses, they have taken on a connotation of being green—cool and fresh.

Some categories of fragrances can't be so easily linked with familiar, naturally occurring substances. For example, aldehydes are special synthetic blends that are frequently used as top notes in perfumes because they connote "sparkle" or "brilliance," which are admittedly difficult terms to define when applied to odors. However, "Chanel No. 5," first introduced in the early 1920s, is still one of the most popular perfumes ever created, and an aldehyde is its dominant note. Other perfumes have been formulated with aldehyde compounds, and they also cannot be linked to a particular substance in nature. If someone were to ask you what "Chanel No. 5" smells like, you'd search for a word and finally say, "It smells like 'Chanel No. 5.'"

What Odor Preferences Tell Us

Numerous studies have tried to establish odor preferences across cultural lines, between the sexes, and among various personality types. For the most part, men and women alike enjoy simple floral odors, but men state a preference for phlox, orange, and heliotrope. Men tend to ex-press stronger preferences among scents generally considered pleasant; women are more likely to react strongly to odors generally considered unpleasant, which is at least part of the reason why a lover's unpleasant odor can cool a woman's passionate feelings very quickly. Males tend to rate the musky smells higher; women tend to like nutty odors, such as almond.

Cultural preferences certainly exist, and the same odors may have entirely different associations depending on the context. For example, during the first half of the twentieth century, European laundry soaps were scented with citronella oil, and it gained a reputation as a fresh, clean odor. In the United States, however, citronella oil is used as an insect repellent and is associated with hot summer nights.

In Japan and China, strong scents are not favored, and perfumes are used far less than in the West. When scents are added, they tend to be lighter and more subtle. The French are said to like warm, sensuous fragrances, such as jasmine, and the Germans prefer the green smell of pine. Obviously, significant market research goes into creating a new scent, and even finding an appealing name is a challenge. (This leads one to question the rationale behind a perfume named "Poison.")

Personality is another factor that enters both marketing and consumer choice. One study, conducted in what was then West Germany, involved giving women consumers personality tests and then comparing the results with their perfume choices.[1] The researchers concluded that extroverted women tended to choose perfumes dominated by the fresh fragrances, while introverted women preferred the spicy Oriental perfumes. Those who were deemed "emotionally ambivalent," meaning that they tended to experience frequent mood changes and tended to be dreamy—traits that are not necessarily negative, despite the label—tended to choose the flowery fragrances. The women who, according to the personality testing, were labeled "emotionally stable" did not show any significant preference toward any specific fragrance category.

While these research results can't be taken as definitive, you might be able to see some correlations in the personal scents you prefer. What should be most important to you and your partner, however, is that the fragrances you wear in a romantic setting please both of you. You may love the aftershave your mother gave you for your birthday last year, but if your partner doesn't like it, then I wouldn't use it for a romantic evening. The same goes for perfume that may have a sensuous name but that your lover finds overpowering and not the least bit sexy.

Ultimately, the choice of scents you use on your person, as well as in your home, result from trial and error. If you keep experimenting, you and your lover will find the scents you enjoy, perhaps even a scent that you share. Since the effect of an odor may change with familiarity, it's important to vary the fragrances you use.

In social and business settings, you will also be identified with the scents you wear, so it's important to choose fragrances carefully. Our individual chemistry interacts with perfumes, and what smells great on one person may not be so pleasant on another person. For women, the way a scent interacts will change during the menstrual cycle and pregnancy. Diet also influences body odor and, therefore, the chemical interaction of perfume on the skin.

When you choose a scent, regardless of its purpose, perfumers recommend that you spray the scent either on your wrist or in the bend of your arm. A bottled perfume can be tested by applying it with a cotton ball. Wait for thirty seconds or so to sniff the odor the first time and sniff it again over a period of several hours—after you have left the store. You can't evaluate a scent in a department store where so many odors compete, nor can you judge it when it is competing with other odors you are wearing. Ask your lover to smell the odor, too; if both of you like it, and your partner finds it romantic and sensual, then you've discovered a bedroom odor, one you will begin to associate with sex. Favorite odors can be "layered," meaning that soaps, lotions, and other cosmetics with the same odor are used simultaneously.

Some people prefer sexual encounters without artificial smells masking the natural odors. This is an issue that should be discussed, not assumed. As our culture becomes more "odorized," it is likely that perfumes and colognes will become an increasing part of a sensual environment. Nothing is unnatural about this; using fragrances to please our lovers and ourselves is deeply embedded in culture. Enjoy the range of choices available; have fun with perfumes and colognes, and make them a special part of your romantic life.

10

Popular Aromatherapy

THE WIDE AVAILABILITY of dozens of essential oils for personal use is a relatively recent development in our culture. We've had vast amounts of scented hygiene products and perfumes to choose from, but now we have the essential oils derived from plants more widely available than ever before. Before I discuss these oils in detail, you should be aware that the vast majority of scents, natural or artificial, have never been subjected to scientific scrutiny. So, if you go to a specialty store or health food market and you see vials of oils for sale for use as "aromatherapy," consider that, as yet, little or no scientific evidence proves that these scents are beneficial to your health in any significant way. They may make life pleasant, and you may find them sensuous additions to the bedroom—and other areas of your home— and they may subtly alter your mood because your reaction to them is positive. What we like often lifts our mood, so I'm certainly not saying they are without value.

Of course, a few specific odors discussed in popular aromatherapy literature are being tested for their effects on the brain. For example, in folklore from various cultures lavender is considered one of the primary "bedroom" odors. But even in popular aromatherapy literature, lavender is said to create a sensuous mood because it is relaxing and soothing, not only because it may cause sexual arousal. When lavender was tested in various research projects, it was found to affect brain wave activity and does indeed have a relaxing effect and promotes sound sleep among the elderly, for whom insomnia is often a problem.[1] In our study of male sexual arousal, lavender was one of the odors that increased blood flow to the penis. In this case, folklore and science found common ground.

At our foundation we've chosen to test certain odors because they are popular and because claims about their beneficial effects appear cross-culturally. Many are mentioned in mythology, sacred literature, and cultural stories and customs throughout the world. When something exists across time and in so many cultures, that usually provides at least a hint of its efficacy. So, even if scientific proof is still lacking, there's no particular reason not to experiment with a variety of scents in your bedroom and find out which ones add "flavor" to your sex life. With some experimentation, you'll learn which odors you enjoy and which may detract from your sensuous mood. You've probably already used popular aromatherapy in the form of scented massage oils or odorized candles.

When I use the term "popular aromatherapy," I'm not talking about actual medical therapies. We are at the beginning of the development of true aromatherapy for health problems and conditions. I'm talking about the kind of aromatherapy that is written about in popular books and practiced in some quarters within the world of alternative health care. Again, I'm not saying that popular aromatherapy is without value; what I am saying is that we don't have scientific proof of its efficacy. In general, the mixtures of oils that are mentioned in the dozens of books about popular aromatherapy are generally harmless, and much of what they recommend is quite pleasant.

From Where Do the Scents Used in Popular Aromatherapy Come?

The scents used in popular aromatherapy come from oils found in plants, and these substances are called essential oils. Such oils occur in small amounts and are found in the bark, roots, flowers, and leaves of many plants. They are also found in the rind of fruits and in the plant resin. These oils circulate within the plant and are concentrated in different locations, depending on the time of day and other factors. The oils provide the scents that attract pollinating insects or keep predator insects away. Aside from being pleasurable substances for humans throughout the ages, these oils play a role in maintaining the health of our planet's foliage.

Essential oils are, by definition, concentrated and should not be taken internally, and in most cases, they shouldn't be applied directly to the skin. Scented massage oil, for example, is a base oil, such as safflower, corn, or almond oil, to which a *small amount* of concentrated essential oil has been added.

Essential oils can be diffused as odors in a room or added to bathwater or to unscented oils to give them a pleasant scent. In most aromatherapy "recipes," the amount of fragrant oil used is measured in drops. Essential oils are generally quite expensive, which is one reason they aren't found in most products marketed for mass use. When you purchase lemon-scented laundry detergent or pine-scented candles, you are buying a product that is infused with a manufactured odor. The vast majority of odors found in commercial products are artificial smells. The scents you use in your bedroom need not be these natural oils either, and many of the aromas mentioned in this book are available in manufactured form. Health food markets, mail order catalogs, and specialty stores usually carry many lines of these prepared oils and products.

The odors we use at our foundation for research purposes are artificial, which is the way it must be in order to achieve accurate results. Essential oils are by nature unstable, and the concentrations vary greatly. In other words, essential oils derived directly from plants are not uniform in their chemistry, so using them for scientific testing is not practical.

Essential oils are extracted from plants by a variety of methods and then sold in very small amounts. Some oils are sold in mixtures, usually determined by what psychological effect each oil is said to have. For example, oil of rosemary and oil of peppermint may be mixed to form a product advertised as promoting an alert, but calm, mood.

Bringing These Oils Home

Other than adding the oils to bathwater or massage oil, the easiest way to use fragrant oils is by diffusing the odor throughout the room. You can do this with an odor diffuser that fits over a lightbulb, in potpourri burners, or in electric odor diffusers that create a fine, odorized mist that is sprayed throughout the room. You can also add a few drops of the oil to the melted wax in an unscented candle. If you are so inclined, you can make sachets and your own candles. You can also easily scent clothing with the oils by putting a drop or two on a cotton ball, and when the oil has thoroughly soaked through, you can place it in a drawer, being careful not to let the oil directly touch your clothing or lingerie.

The important thing is to create the mood you enjoy and let scents help you do so. Begin with a playful mood and have some fun trying odors you may not have used in your bedroom before. You have an abundance of scents to choose from, and it's not surprising that some of the most popular odors are those that appear in folklore of romance.

A Rose Is a Rose Is a Powerful Symbol

Some women say that a man with a bunch of roses in his hands is about as irresistible as a man can get. In many myths, roses are the symbol of true and enduring love, and in aromatherapy lore, the scent of roses is considered an intoxicating scent, rather than a relaxing odor or a fresh fragrance. We know that in ancient Greece, Rome, and Egypt, rose oil and rose water were used liberally among the classes that had servants to extract the oil and prepare the products.

Cleopatra certainly believed roses have aphrodisiac properties, and she had attendants drench the sails of her barge with rose water before her first meeting with Mark Antony. (She used a host of other scents, too, and from what we can gather, her barge was a floating perfume garden.) According to the Sufis, the rose is considered a symbol of the path to God, and ancient poets from the Middle East praised the rose as the greatest symbol of love. In our culture, roses are sometimes given early in a romance as a way of saying that these new feelings are serious and not a flight of fancy; as the years go by, roses are a way of showing that love is still blooming.

In addition to bringing the gift of roses into the bedroom, where their fragrance can fill the air, you can also use rose oil and rose water in the bath. We can't say that the scent of roses is actually a sexual scent, but we know that the symbolism of roses is powerful. Some people like to sprinkle rose petals on their sheets before they make love so they can inhale the fragrance and so the odorous residue clings to their skin. The silky texture is also appealing. Rose sachets give women's lingerie a pleasant scent but a little rose scent goes a long way, and it can be cloyingly sweet if used to excess. When any scent is too strong, it becomes the focus and takes attention away from other elements of romance. So I recommend using rose oil—and most scents—sparingly.

A woman I interviewed said that she loved getting roses and she enjoyed their scent, but she considered the smell more conducive to sentimentality than sex, so she doesn't care to have the odor in the bedroom itself. Her reaction is probably true for many odors that have great symbolic value. Experimenting is the only way to learn about your individual responses to fragrances.

The Queen of Sexy Odors Is the King of Flowers

In aromatherapy literature, jasmine is said to work its magic on both men and women; since ancient times it has been used in spells and love potions designed to "capture" the heart or rekindle fading romance. Jasmine is commonly used in bridal wreaths in India and, like the rose, is symbolic of deep and lasting love. In the *Kama Sutra*, a man is advised to create the atmosphere for love this way:

> Give her your jasmine garland
> lie her back gently
> and massage her body with sweet sandal-oil.

Many massage oils marketed today feature jasmine or sandalwood, and both scents are said to promote relaxation. The fact that jasmine releases its fragrance at night no doubt adds to its exotic reputation.

In much folklore, jasmine is considered an odor that promotes confidence and strength, as well as pleasant relaxation. It is also said to soothe tight, aching muscles and sore joints after a workout, and for this reason, it's often added to massage oils. Jasmine promotes wakeful relaxation, as opposed to a drowsy mood, making it one of the erotic scents in popular aromatherapy literature.

Many references to the fragrance of sandalwood oil occur in the lore about the sexual power of odors. Like jasmine, it is pleasing to both women and men, and for this reason, it may be a good candidate for a shared scent. According to the legends about sandalwood, its woodsy smell has the power to distract us from our worries and lure us toward the earthy, sensuous pleasures. Because of this ability, it is frequently used to promote a meditative state, which by definition means that the chatter of the mind slows down. Given its wide use and appeal, it appears that our ancestors had to find ways to combat the stress of everyday concerns just as we do.

Strength, Confidence, and Sexual Vitality

Juniper is a common scent that is said to promote strength and confidence, perhaps because in folklore it is considered the plant most associated with protection and cleansing. Once considered an important spiritual plant in Europe, juniper was said to protect against witches, and juniper wood was burned to banish demons. Its reputation as an erotic plant comes from its purported ability to restore strength and ease emotional burdens. It is also said to relieve negativity we bring from past relationships and, thus, enable new lovers to build trust.

From what we can gather from various myths and stories, it appears that using smells to promote a relaxed mood is only a part of the story. Odors that stimulate and invigorate us are just as important in the folklore of fragrances.

Interestingly, ginger appears widely in stories and myths about sexual performance. In fact, the Persian physician Avicenna advocated a mixture of honey and ginger to "cure" male impotence. This physician of old stated that ginger increased blood flow to the penis and, hence, restored "manly" erections. Apparently, a similar passion "recipe" appears in Turkish folklore, and it claims to lead couples to burning desire while simultaneously promoting fertility.

Ancient Romans sometimes combined ginger, whose oil is distilled from the root, with rhubarb, cinnamon, and vanilla to arouse sexual passions during festivals. Ginger has a reputation throughout Asia and the Mediterranean of being a powerful aphrodisiac, the very reason English Puritans disliked it and discouraged its use. Ginger's erotic reputation and the ancient physician's "clinical" observations are especially interesting because, as I've noted, pumpkin pie is often spiced with ginger—and cinnamon, too—which our twentieth-century research found had a physiological effect on male arousal.

When people say they want to "spice up" their sex lives, the way to do so might quite literally be found in the spice cabinet. In popular literature about the sexual power of odors, spices are frequently mentioned, including one so common—black pepper—that we wonder why claims about its ability to enhance our sexuality aren't more widely known.

Black pepper, whose oil is extracted from the plant's berries, has a reputation for being a very powerful sexual scent, perhaps because it is considered a stimulant. In sexual mythology, pepper is one of the scents used to "wake up" desire between partners when ardor has cooled. In fact, many of the aphrodisiacs and scents considered erotic were recommended as aids to invigorate sexuality. Even in ancient times, it was known that familiarity between partners could lead to a gradual waning of passion, and certain fragrances and foods were recommended to keep desire alive. Certain scents were specifically used to arouse interest in a partner when time and familiarity have taken their toll on passion. This situation sounds exactly like the subject of numerous articles and books today, but somehow we've been led to believe that it is a uniquely modern-day problem. In the past, the solution was seen in stimulating the senses, where today, we tend to see the problem and the solution as based in psychological or even sociological issues. Given what we already know about the power of odors to alter mood state, perhaps we should give the senses another chance!

In aromatherapy folklore, black pepper is said to be especially useful for men, and the oil was used to increase stamina on the battlefield as well as to build stamina in the bedroom. This spice, as common today as salt, was once considered as valuable as gold.

Basil is another common kitchen spice with rich folklore surrounding it. Its sweet and spicy aroma is associated with fertility and seduction. Legend has it that, in Italy, young women used basil to "bewitch" the young men of their choosing, and if the powers of the love goddess Venus were needed, basil was the agent used to call her forth. In other words, basil could arouse sexual interest and boost libido.

An Orange Isn't Just an Orange

A fragrance known as neroli, distilled from orange blossoms, is associated in ancient myths with both sexual purity and seduction. That explains why orange blossoms have been used in bridal wreaths as a young woman's way of proclaiming her virginity. On the other hand, the scent was also favored by prostitutes in Madrid because of its power to seduce.

The oil of the orange itself is also considered to have an invigorating odor and adds zest to the bedroom. In Roman mythology, the orange is the "golden apple" that Juno gave to Juniper when they were joined in their "celestial wedding."

Like orange oil, the essence of lime is said to energize, and it is supposed to add laughter and playfulness to the atmosphere, too. In aromatherapy folklore, lime oil has the ability to relieve anxiety and clear the mental pathway for communication between lovers. In myths, the image of two side-by-side lime trees represents the perfect marriage of two individuals who both blossom and bear fruit.

As I've noted, in the West citrus odors are often added to products to create a fresh, light smell that people associate with cleanliness. Lemon is the most commonly added odor, and as far as I know, it does not carry the same erotic connotation as orange and lime.

A woman I interviewed said that orange scent adds to the erotic atmosphere when she and her partner use it in the bedroom during the day. She explained this by saying that she associated citrus odors with morning sex and breakfast in bed, but at night, she and her partner use fragrances that she considers more intoxicating and dreamy.

The evening fragrances this woman and her partner enjoy are a combination of the spicy and woodsy smells, with the nutty odor of clary sage added, too. Lavender is favored when this couple want to create a romantic atmosphere, but the amount used in an odor diffuser was, as she said, "cautiously small," because when it was too strong, it made them sleepy.

In talking to this woman, I was struck by the variety of moods she and her lover created. "In the morning, we like the freshness of the citrus smells, and we prefer to let natural light into the room," she said.

143

"Sex wakes us up and sets a pleasant tone for our day. Some evenings our sex is more 'athletic' and the mood isn't so romantic as it is lustful and experimental. So the odors we add are more spicy and stimulating. On other evenings, we are in a more romantic mood, and we spend more time focused on each other. We might take a glass of wine to bed or give each other a massage. We light candles, and we scent the room with lavender. The mood is dreamy, and we fall asleep easily. So, we choose the smells we use with care—and my partner also likes to experiment."

I recommend taking this woman's experience to heart. What do you have to lose? With the array of odorized products and essential oils available to us today, we can regain knowledge that ancient people took for granted.

A Garden of Fragrances

The scents listed above are only a few of the most popular smells that we find in folklore and in popular aromatherapy. Others include:

CEDARWOOD: In the Bible, cedarwood is referred to as the "tree of life," and in other legends, it's called the "tree of the gods." It's considered useful for relieving fear and anxiety, but in folklore, it's also linked with fertility.

GERANIUM: Most of us think of geranium as a cheerful flower that brightens our porches and decks. And so it is. In folklore, it is also considered a plant that most perfectly balances male and female energies. In Chinese philosophy, it represents the balance of yin and yang energies and, therefore, promotes harmony. For this reason, it's said to have the power to smooth differences between lovers and to promote equilibrium.

PATCHOULI: You may see patchouli commonly used in incense products for the bedroom because in popular aromatherapy it is used to provoke sexual energies and help us overcome inhibitions. Patchouli became widely known during the sixties and seventies and was considered an exotic smell brought to us from the Eastern world. In many circles it still is considered quite exotic, which can add mystery to the bedroom.

YLANG-YLANG: A honeymoon symbol in Indonesia, the scent of ylang-ylang flowers is purported to be an aphrodisiac and to promote highly erotic and sensual sexual experiences.

VANILLA: In folklore, vanilla is favored by older men, which was a finding of our study of male sexual arousal. We don't know why this is true, but that doesn't matter. Earlier in this century, psychoanalyst and sex researcher Havelock Ellis wrote that workers in a vanilla processing plant were in a constant state of sexual arousal because of their exposure to the scent. We don't know if that's true, but in surveys, vanilla is considered one of the fresh, clean smells and probably lightens mood because it is so universally pleasing. Go ahead and add vanilla to your bedroom and see what happens!

FRANKINCENSE: We know this as a prized gift that the wise men from Egypt were said to have brought to the infant Jesus. The plant actually has a rich, spiritual history starting long before that event, however, and it was long considered an odor that transformed consciousness. In other words, it has the ability to alter moods, and for purposes of sensuality, proponents claim that the inner peace and calm it promotes can lead to a relaxed, romantic atmosphere. The plant also contains musk, which of course is associated with sex.

MYRRH: According to some legends, myrrh is the oldest perfume, and women wore small bags with myrrh around their necks so that their body heat would release the seductive odor. We also know it was highly praised as a sacred scent and was another of the gifts said to be presented to Jesus at his birth. Along with frankincense, myrrh is usually used in combination with other odors and is sometimes used in massage oils. Myrrh also contains musk.

Being familiar with these odors and the stories around them can be a useful place to start when you begin adding odors to your home, including scents you use to odorize your bed linen and lingerie and, of course, your body. In our culture, much of the folklore and mythology about odors and their power has been lost, but we are entering a new era, one in which the nose is again assuming an important and prominent place.

As I said in the beginning, I'm presenting this information even though it isn't scientifically verified. For recreational use in the bedroom that doesn't matter. But as we learn more about individual odors, we may find that at least some do act directly on the brain and change our mood. The pumpkin pie and lavender combination and the cinnamon buns certainly lead us right back to the folklore of the past.

If you want to spice up your sex life, then by all means start adding some of these special scents to your environment. I don't think it matters if they are artificial smells added to products or if you buy the essential oils and use them to add odors to your lingerie or linens or in scent diffusers.

Again, I recommend adding these odors in small amounts because any strong, concentrated odor can not only be unpleasant, it can cause nausea or a headache, too. Obviously, you don't want that. Anyone with asthma must be particularly cautious about odors, and some people have allergies to certain plants. So, use your common sense and enjoy the common scents that may enhance your sexual pleasure.

11

"Now I'm in the Mood"

YOU CAN FILL your bedroom with sensual fragrances, watch erotic movies, and play your favorite romantic music but have all your efforts be for naught if you're angry, preoccupied with concerns about work or family problems, or feeling down over an event in your life. It should go without saying that you must be in the right mood to enjoy sexual relations, but sometimes you may forget this important element of lovemaking.

Of course, the critical questions are: What is the "right" mood? How do you achieve this mood? Can you change your mood—at least some of the time?

Odors Aren't the Whole Story

Every couple go through periods in their relationship when sex becomes unimportant or even out of the question because of other pressing concerns. Usually, both partners know when they are in these stress-filled, often chaotic times. A lack of understanding and sensitivity about a partner's needs or concerns could have serious consequences for the relationship.

Emily, a woman I interviewed, ended her marriage because her husband expected their sex life to continue as normal even though she was going through a family emergency. (This was obviously not the

147

only problem, but their sex life became a setting from which deeper problems emerged.) Emily's elderly parents were both gravely ill one spring. Her father was in a rehab center, unable to speak or move as a result of a stroke; her mother, who was going through chemotherapy, came to stay with Emily and her family. Emily took a leave from her job and spent all her time taking care of her mother, while regularly visiting her father and attempting to keep his spirits up as he struggled through rehab.

"This was an unusual—and horrible—situation," Emily said, "and I knew it wouldn't last forever. But for obvious reasons I was concerned about my parents and tired from tending to their needs, not to mention taking care of our own children. I just didn't feel romantic or sexual. What I needed from my husband during those weeks was help and support—I needed a friend. I didn't need selfish demands from a man who pretended like nothing was different in our lives. He acted like a small child and was angry because I wasn't paying enough attention to him—that meant sex. I faced things about him I hadn't had the courage to look at before, and I filed for divorce after the crisis passed and I had the opportunity to sort out my feelings. The hardest realization was that I didn't *like* him. I just couldn't let life go back to the way things had been before our crisis. I didn't want to spend the rest of my life with such a selfish man."

Obviously, Emily's situation is extreme, but it illustrates the difficulties long-term relationships may encounter, and often sexual problems are only a symptom of a deeper schism. If this couple had been emotionally closer, the husband would have pitched in and helped out, and he certainly wouldn't have made sexual approaches to his wife. He would have realized that she couldn't possibly be receptive to sex when the mental and physical demands on her were so great. In this case, mood isn't even the issue.

Most situations aren't as difficult as Emily's, however. More typically, our moods not only shift day to day, but even fluctuate hour to hour. We may wake up feeling great, but stress and various negative incidents leave us frazzled and perhaps even anxious and tense by evening. If we find this happening day in and day out and we aren't much fun to be around at home, then we need to take a look at ourselves and our lives.

While lack of sexual desire is the main difficulty of almost half of those who visit sexual dysfunction clinics, bad timing on the part of their partners was the major complaint voiced by women in a study of one hundred couples.[1] Another survey reported that, for the majority of men, fatigue was the biggest barrier to sex, with lack of time coming in second.[2] These results were essentially the same for men of all ages, although men age forty-five and older tended to rank their partners' lack of interest as a more important issue than lack of time.

While I don't wish to engage in excessive stereotyping, men and women tend to respond differently to sexual cues, and lack of interest in sex may come from different causes. For a woman, lack of interest in sex is often related to what's going on in her life and the quality of emotional intimacy she shares with her partner. This is what Emily expected her husband to understand. In numerous books about male sexuality, it's emphasized repeatedly that, for women, "foreplay" takes place outside the bedroom. To many women, lovemaking—including foreplay—is defined as a man showing genuine interest in the woman's concerns, opinions, ideas, and responsibilities. The behavior that precedes sexual encounters is at least as important as what takes place when the lights are dimmed and the clothes are in a pile on the floor. Emily was turned off to sex with her husband when he revealed himself as a shallow, selfish person who had little regard for what was happening in their family. Emily told me that she once thought of her husband as an attractive, appealing man, but his behavior was ultimately more important than his looks, and he lost his sex appeal—fast.

As women's roles have changed and most women have full lives outside the home, men's lives have changed, too. Unfortunately, many men are resistant to the idea that their home life must be characterized by a cooperative spirit and that their partners expect to be treated as full equals with their needs—including sexual needs—respected. Certainly, many men have made adjustments and prefer the world of gender equality. And of course, throughout history, millions of men and women have been caring lovers to each other regardless of the roles they were expected to play. I mention this here because a successful sexual relationship depends on the atmosphere in our homes and the general mood of the partners.

While we know that odors can't solve all fundamental relationship problems, various scents *do* have an effect on our moods and sexual feelings. Our sense of smell can help us live more comfortably and pleasantly in many ways. Certainly, odors can subtly shift our moods, and they can enhance our sex lives because odors we like tend to lift our mood.

Various fragrances can also help us relax and quite literally "spice up" our sex lives. Adding odors to our environment provides variety and stimulates our senses, perhaps distracting us from small, unimportant concerns and helping us focus on sensual pleasures. It seems to me that our society has focused on the visual and tactile components of sex but has not celebrated the total sensual experience. We certainly haven't adequately explored odors and how they influence our moods and emotional states.

What We Have Learned in the Lab

When our Smell and Taste Research Foundation began studying odors and their effects on the brain and mood, our approach was medical, and we were looking for safe, noninvasive treatments for common conditions and disorders. For example, when we learned that the odor of green apple can reduce anxiety, this had important implications for the medical treatment of people who are diagnosed with generalized anxiety disorder. Lavender and vanilla have a similar effect, meaning that these scents have potential as legitimate treatments.

Lavender, for instance, acts directly on the brain by enhancing alpha brain wave activity in the back of the brain. When there is an activation of alpha waves, we generally are more relaxed. When individuals peacefully meditate with their eyes closed, the alpha brain wave activity increases, and they experience a sense of deep relaxation. Achieving this relaxed state is often the primary motivation for beginning the practice of meditation. With practice, meditation is the easiest, most reliable way to achieve inner calm. It offers other physical and psychological benefits, too, the by-products of the relaxed state that people who meditate regularly experience.

From the Lab to the Living Room

Most people feel anxious once in a while and may bring that anxiety home from the office. Anxiety and sex don't make good bedfellows, so to speak, and given what we know about green apple, for example, if we're anxious, it makes sense to use the scent in our homes in the form of a scented candle or a room spray. You needn't be diagnosed with generalized anxiety to take advantage of the research.

Peppermint is another scent that may offer benefits in our homes because it tends to be a "wake-up" scent, but it doesn't make us nervous or "hyper." An airline in New Zealand offers a peppermint odor to their first-class cabin passengers at the end of transcontinental flights to help them wake up and have a positive feeling about their trip. If they've been sleeping, the peppermint odor induces an alert mood state and takes the edge off the groggy feeling so many passengers experience after a long flight. An alarm clock manufactured by a Japanese company emits a peppermint odor ten minutes before the alarm sounds, which helps the sleeper move from a sleeping to waking state quickly and pleasantly.

If you come home from work and feel tired out, you have several options. You could take a quick nap before dinner, but many working adults don't have that luxury. Besides, naps late in the day tend to disrupt nighttime sleep and aren't a good idea for most people. (Early afternoon naps are beneficial, but few of us are able to indulge in that pleasure.) You could splash cold water on your face—not a pleasant experience, nor necessarily effective. Or you could inhale the scent of peppermint, an odor most people find pleasing. Inhaling this odor may be just what you need to begin shifting your focus from problems at work to the comfort of your home. An odor can be a way to make the transition you need to enjoy your evening at home with your family.

If fatigue caused by lack of sleep negatively affects your mood, then lavender may help you fall asleep more easily. Many elderly persons report problems with insomnia and other sleep disturbances, and studies have shown that the scent of lavender can promote sound sleep among this population group, and it may help individuals of any age. We also know that the odor of lavender has been a popular bedroom scent for several centuries, and it is associated with relaxation.

With the availability of a wide variety of scents, you can take the information gathered in the lab and brighten up your home environment with odors. While fragrances don't guarantee great sex, they can help you create the mood and atmosphere conducive to emotional intimacy and a relaxed evening. In fact, odors will probably soon become part of home decorating; just as you choose colors and fabric textures carefully, you may choose scents to help create a desired atmosphere.

Awake, Aware, Relaxed

When we stimulate our sense of smell with pleasant odors, we are receptive to cues in our environment. For example, jasmine, like peppermint, enhances beta waves in the front of brain and, hence, tends to stimulate us and increase our awareness of other stimuli around us. Fatigue or tension makes us less sensitive to sensory stimulation of any kind, but an alert state allows us to notice more stimuli, including sexual cues.

Relaxed doesn't necessarily mean sleepy, although in our hectic culture, many of us tend to link the two. Relaxing with a good book may mean falling sleep after a few pages. But relaxation can also mean an alert, awake, peaceful mood, the kind of mental state that allows us to notice what our partner is wearing and be aroused by the scent of colognes or other fragrances. It may be the kind of mood we experience at a party where conversation with friends is pleasant and intimate. We notice the aromas coming from the kitchen, and the variety of sights, sounds—and smells—leaves us pleasantly relaxed. This relaxed state may make us receptive to intimacy and romance. If you want to enhance your sexual and romantic life, then you want to create this kind of mood.

Variety Is the Spice

The frequency of sexual relations tends to decrease if couples have sex the same way all the time. Boredom with sex is the most frequent complaint among happily married men I interviewed. The obvious solution is not to seek new partners, but rather to vary the place and time of lovemaking with their wives, as well as experiment with different sexual techniques.

Odors tend to be less effective, too, when we are exposed to them over a long period of time. If you use the same odor in your living room day after day, the effect will eventually diminish, and you may not even notice the odor. The same is true of perfumes and odorized lotions, soaps, and colognes we use on our bodies and for the scented products we use in our homes. Experiment with a variety of odors. If you or your partner don't like a particular odor, then try another. If a particular odor makes you so relaxed that all you care about is sleep, then avoid that odor until you're ready for a good night's rest.

While certain scents are popular and traditionally linked with sensual pleasures, reactions to odors are by no means universal. For example, if you take what you read in this book to heart, you might buy some lavender-scented products for the bedroom and fill your living room with the odor of jasmine. Your partner notices the jasmine and responds positively, but the lavender in the bedroom is a turn-off. Maybe it reminds your partner of another person, one with whom he or she has had a rocky relationship in the past. All your partner can think about is that person, and now the mood is changed. The reason doesn't matter—try another fragrance, even if you like the odor.

Vary the odors you use to scent lingerie, bed linen, and your body, just as you vary other elements in your lovemaking. With the number of odorized products, perfumes, and essential oils available to you, changing scents shouldn't be difficult, prohibitively expensive, or time consuming.

Back to the Brain

Our moods are dependent on many factors and change numerous times during a typical day. We may become annoyed when we can't find a parking place, pleasantly surprised by an incident at the office, joyous when we receive a bonus, angry when a relative calls to borrow money, disgusted by an item on the news, pleased by the taste of dinner, and alarmed by our credit card statement. We may relive all these incidents and experience new feelings as we think about our day. Feelings come and go, and most people are able to change their focus and put some concerns aside. In fact, good love relationships are dependent on our ability to adjust to the events in our lives and shift our focus. A bad day at the office doesn't have to be brought home.

Odors can't control or create all your moods, but they can help you change what you focus on and how you feel. A good food smell can distract you from an angry thought, and soon your attention is redirected to what is simmering on the stove. Likewise, the smell of vanilla or jasmine in your living room may perk you up and redirect your concentration away from thinking about how tired or stressed you are. Odors, then, become one of the external stimuli that allow all your moods to flow along, changing moment by moment.

Don't take my word for this—use your home and your bedroom as your personal laboratory. Use what we've learned about odors and let them add pleasure to your life. Nothing enhances relationships more than creating an atmosphere that helps us focus on our partners and the sensual delights of romance and sex. Let a variety of fragrances help you create a sensually pleasing home and a sexy bedroom.

12

A Symphony of the Senses

THUS FAR, this book has focused on the significance of the sense of smell in the arena of sexual desire and erotic expression. If you use the ideas in this book, you will add variety, a sense of adventure, and a playful spirit to your sexual encounters. When you use odors, you may also intensify the sensations and pleasure of your experiences. When it comes to the total experience of sex, however, your sense of smell can't be isolated from the overall quality of your romantic relationships, your fundamental sexual attitudes, and your awareness of your other senses. All your senses work together to create a positive sexual experience, and now you can widen your explorations of sexuality to include this new information about odors and sex.

The most emotionally satisfying lovemaking is the result of healthy attitudes and your appreciation of the variety of sensory stimulation. This presupposes that the desire for lovemaking is mutual and a fundamental attraction brought you and your partner together in the first place. Once the basic psychological and emotional climate is present, you can enjoy all your senses, each of which has a special erotic function.

Sensuous Touch

To imagine a satisfying sexual life without actually touching your lover is impossible. While individuals exist whose sexual gratification comes almost exclusively from watching sexual activity, they comprise a very small minority of the population. For most adults, touch is essential, and during lovemaking, it is usually perceived as the dominant sense. The other senses may bring us together, provide stimulation, and bring their own unique delights, but touch is primary in our sensations of sexual pleasures.

Touch is a fundamental form of communication, and beginning in infancy, our bonds with other people were formed through touch. Touch is so important to infant development that lack of adequate touch is associated with *marasmus*, which means "failure to thrive." Once thought to be the result of inadequate nutrition, inadequate touch as part of the cause was documented when the high infant death rate in orphanages was studied. Numerous animal studies confirm that warm, caring touch is necessary for normal physical and psychological development.

As you matured from infancy to childhood, you learned that certain kinds of touch are appropriate—or inappropriate—in a variety of situations. You learned that hugs are usually reserved for close relatives or family friends and that embracing strangers is inappropriate nearly all the time. When you were an infant and a small child, you were uninhibited about touching yourself; eventually you learned that it isn't acceptable to touch your genitals in public. Although there is great variation in attitude about touching oneself, some children are taught that it's never okay to touch their genitals, even in private, especially during adolescence.

By the time we're adults, most of us have absorbed information about the various levels of touch that we use in our culture. This is a basic component of socialization. Through observation, we've learned that in our culture, the handshake is an acceptable form of touch in both business and social settings. We know that Aunt Sally likes a warm embrace when we visit her, but Aunt Grace is more reserved, so we keep our physical distance.

In our culture, a hug between two women is viewed as a warm expression of friendship. Women can touch each other's hands or kiss each other on the cheek, and an observer would simply think that they are close friends or relatives. Among most men, however, a slap on the back or a hearty handshake has been, until recently anyway, just about the only acceptable expression of affection. (Men in other cultures, particularly those in Eastern Europe and the Mediterranean countries, often greet one another with cheek kisses or a hug. However, these practices tended to remain part of private behavior when people emigrated to this country, which from its beginnings was influenced by the less-touch-oriented northern Europeans.) In some families, male relatives—even fathers and sons—remain physically distant, and they consider exchanging a hug "unmanly." Fortunately, these barriers have broken down in the last few decades, and more men are comfortable touching each other in physically affectionate ways. This is a positive development, because many men grew up associating touch only with sex, and they received most of their touch in sexual settings. This is one of the reasons some men have difficulty expressing nonsexual affection, even with their lovers.

Sexual touch is a sensitive subject, because it is the most intimate type of touch. Erotic touch is generally a private activity, and mutual consent is essential. If erotic touch isn't consensual, it becomes threatening and potentially very damaging; we all know that great psychological damage may be done when children, for example, are touched by strangers or family members in a sexual way. In fact, when that happens, we no longer call it touch—it becomes molestation or abuse because it involves wielding power and using coercion or force. Sexual harassment, sexual assault, and rape are crimes of aggression and power carried out in a sexual context.

Touch and Intimacy

Within adult relationships, and adolescent bonds, too, sexual touching usually moves through levels that are determined by and based on mutual trust. Caressing another person's face, for example, is a powerfully intimate touch, as is touching the leg or torso. In a sexual context, our hands become a vehicle of intimacy, and holding your partner's hand, while not necessarily sexual, is intimate and affectionate. Among adolescents—and the older set as well—holding hands in public is a symbol that the relationship is no longer a simple friendship, but is now a romance. Kissing the face indicates one level of intimacy; kissing the mouth usually means that the relationship is becoming closer.

What is often forgotten is that sexual encounters with our regular partners may need to progress through these stages of touch. Women are often quite outspoken about the need for seductive and tender touching as part of arousal. A woman I interviewed said that she is nostalgic for her adolescent dating days, when holding hands and enjoying lingering kisses on the front porch aroused powerful sexual feelings. Other women have told me that loving, affectionate touch, including kissing, is often forgotten as the years of marriage pass.

The Intimate Kiss

Sheik Nafrazzi, the author of *The Perfumed Garden*, was quite specific about the value of kissing in erotic life when he wrote: "That which is to be looked for in coition, the crowning point of it, is the enjoyment, the embrace, the kisses. This is the distinction between the coitus of men and that of animals." More graphically, he wrote, "The languishing eye puts in connection soul with soul, and the tender kiss takes the message from member to vulva."

Kissing is one of the most intimate erotic touches, because our mouths are associated with nurturing and, of course, with expression. According to a marriage counseling organization in Great Britain, couples tend to stop kissing each other before they stop having sexual intercourse. Many prostitutes will not allow their clients to kiss them because of the symbolic intimacy of kissing. A prostitute will kiss her own lover, of course, but not the strangers with whom she has a business arrangement. Of course, kissing is an important vehicle for inhaling the scent of a sexual partner, which could be an additional reason prostitutes don't want to engage in kissing. If the client's scent is a powerful turn-off, the business transaction could be difficult to perform. Conversely, if the client's scent is a turn-on, then the objectification of the sex act could be compromised, too. This may be an *unconscious* reason that kissing is avoided in impersonal sexual encounters.

Because intimate touch often involves the so-called erogenous zones, you may forget that the whole body is potentially erotic. So, while the genitals are obvious "hot spots," remember that kissing or caressing the face may be the initial erotic gesture that leads to arousal. The inner thighs are sexually sensitive, but stroking the upper arm may be preferred in the initial stages of lovemaking. As the body becomes aroused, the genitals and other highly erotic areas of the body become increasingly sensitized, which means that slowly progressing through the levels of touch intensifies overall sensation and pleasure.

The Olfactory Kiss

No discussion of kissing would be complete without bringing in the issue of the olfactory kiss. In our culture, children are taught about the

"Eskimo kiss," which is essentially rubbing noses. We've tended to think that this is an oddity, and that's why we use rubbing noses with children as a form of affectionate play. But the olfactory kiss, which is essentially using the nose as a greeting or as a sign of affection, has been documented throughout the world. In some South Sea Island cultures, touching one's nose to another person's face, nose, or even the hand is a salutation, much like shaking hands or cheek kissing. What we know as kissing is probably a development that arose as much from the desire to smell one's partner as to touch him or her. Remember, too, that pheromones may be detected in saliva, so kissing involves a pheromonal exchange both through the nose and the mouth. As we all know, kissing and inhaling the scent of our lover are both erotic.

Expanding Touch

Sexual touch can be erotic through clothing, and the feel of the body through the texture of fabric is sexually exciting for both partners. A man I interviewed said that he and his wife like to undress each other, and they choose the clothing they wear for their nights together for their erotic value. Visual cues may be involved, too, but for this couple, touching through clothing before skin caressing is a major component of their foreplay.

Uninhibited and open about the kinds of touch they enjoy, this couple often includes massage in their sexual life, which is a way to nurture and focus attention on each other. They choose massage oils with scents they find arousing, and they vary the setting for their massage and lovemaking as much as possible. When they're both in an especially playful mood, they rub each other with ice cubes or tickle each other with feathers, which adds even more tactile sensations.

While most sexual touch involves the hands, the body, and the genitals, you aren't limited in tactile erotic sensations. If the feel of satin sheets or a silk robe or the gentle tickle of feathers arouses you and enhances the sexual experience, use it. Variety really is the spice of erotic life.

Sexual boredom is often a result of a reluctance to try something new, and many men and women are embarrassed about disclosing their sexual inhibitions. In our culture, eroticized as it is, some individuals feel guilty about having too much fun during sex! In fact, if they step outside the bounds of what appears to be conventionally sanctioned, then they believe they must be doing something wrong. In addition, trying something new in bed can make almost everyone feel vulnerable—exposed. But sex is an arena in which the adage "nothing ventured, nothing gained" applies.

The Erotic Eyes

Until we learned important information about our sense of smell, vision was considered the initiating sense of sexual attraction. Visual cues are important, and for a variety of reasons, often having to do with childhood impressions and socialization, we are attracted to certain physical features in our partners. What our research is beginning to show, however, is that it's entirely possible that we are initially attracted to the odor of a person and then we rationalize about his or her appearance. We say we're attracted because the man is tall or the woman has a beautiful smile, but those may be simply secondary impressions. What attracted us in the first place is the person's odor.

What we consider physically attractive is also based on social factors. Women preferring tall men is partially explained by the fact that being tall is considered a positive characteristic for males, one often associated with success and power. Men may be attracted to thin women, who are classically beautiful, because in our society these women are considered the ideal. (Nowadays, tall women are considered part of the ideal, too.) A man's status in society rises if he is able to "catch" such a woman. Obviously, I'm stereotyping here for the sake of simplicity, but while attraction tends to seem quite individualistic, we are influenced by everything from media images to our parents' expectations to our aspirations for status in our society.

When Love Really Is Blind

Numerous stories and movies exist about what happens when a blind person falls in love, and because he or she can't see the other person, the typical visual biases are absent. Race and conventional standards of attractiveness don't interfere, and in these stories, the blind person can "see" the inner beauty of the potential partner without the influence of superficial features. Like Romeo and Juliet, these love affairs are usually "forbidden" and raise the wrath of parents and are considered unacceptable by society. In the 1960s, the movie *A Patch of Blue* told the story of a young, blind woman who fell in love with a very kind and generous man. It was a difficult romance, however, because he was black, and she was white. The theme of the movie centered around her inability to understand the racial taboos and his attempts to explain them to her. Essentially, the movie was telling us that if we were all blind, race would cease to be an important issue in human relationships.

Stimulating the Eyes

It is impossible for most of us to imagine a situation in which we can't see our partner and call up an image of his or her face and body. You may enjoy the way your lover walks, and you notice graceful hand gestures and other mannerisms. Next to touch, visual cues and stimulation are considered essential to erotica. Studies have shown that men tend to be more visually oriented than women, which at least partially

explains men's attraction to erotic art and sexy movies and photographs. To confirm this, all one needs to do is compare the range
of magazines published for the male audience with the dozens of
women's magazines on the market. In both cases, sexuality sells, but
women, more than men, read about sexual issues that relate to their
lives. Men tend to look at the sexually stimulating photographs and
may then proceed to read the articles about automobiles or investments
or perfecting their golf swing.

This is not to say that women are not visually stimulated by erotica, but women tend to be interested in the context or the story in
which the sexual activity takes place. This is why many women aren't
aroused by hard-core pornography but may be aroused by erotic scenes
in movies in which the characters seem real and the plot is suspense-
ful. Even the written erotica produced by and for women usually
includes a story line, and more is left to the reader's imagination.
Sexually stimulating material targeted to a male audience usually leaves
little to the imagination.

One man I spoke with told me that he had almost ruined his marriage because of his interest in what he called "raunchy, even somewhat
violent, porno movies." His wife refused to watch the movies with him
because they were so demeaning to women, and she was uncomfortable that he was aroused by the images and then wanted to have sex
with her. Eventually, no sexual experience with his wife stimulated him
as much as the pornographic images. One day he realized that mastur-
bating while watching a graphic movie had become his only sexual
outlet. Ultimately, he gave up the pornography, because it had become
a serious problem for him, one that interfered with his life and severely
damaged his relationship with his wife.

Most adults are not adversely affected by occasionally viewing
pornographic or erotic movies, but for these images to positively affect
a couple's sex life, both partners must find them enjoyable and

arousing. Throughout the ages, erotic painting and writing have been considered aphrodisiacs. They both stimulate and intensify sexual experiences. Nowadays, we can add film to the erotic repertoire, and for couples, the idea of using erotica as a sexual stimulant is to play out the fantasies on the screen with each other. Certain types of pornography, the kinds we label hard-core, may legitimately be a gender issue, but that doesn't make all erotica wrong or "politically incorrect." If partners communicate well, they can usually find movies that are acceptable to both partners. According to the Adult Video Association, women rent more than half of all erotic videos, so the image of men sneaking in and out of video stores doesn't accurately reflect reality.

One reason some women are uncomfortable with a shared experience of even mild erotica is that they worry about "living up" to the images on the silver screen, where no one has a blemish or a wrinkle. Most women don't look like centerfold models or actresses, who are typically filmed in such a way that any flaws in their body aren't apparent. The actors usually look better than most men, but men tend to ignore that more easily than women, mainly because women are still more harshly judged by their appearance. This may become an even more significant issue as the years pass and signs of age appear. Consider the phrase "she's losing her looks." No equivalent phrase in our language is applied to men. As the baby-boom generation ages, it will be interesting to see if adult movie producers begin casting older actors and actresses and perhaps even changing the themes of the movies to satisfy the changing needs of a very large percentage of the population.

According to the Revlon survey mentioned in chapter 8, 72 percent of women found positive comments about their appearance sexually arousing. So, to male readers: consider gazing at your lover's body *and* telling her how beautiful she looks as part of foreplay.

Lights On or Off?

A couple I interviewed told me that in addition to reading erotic stories aloud to each other in bed and watching steamy movies, usually the more standard Hollywood variety, they also adjusted the lighting to satisfy their differing visual needs. They use candles to create soft light that allowed him to see her body, but without the bright glare of an overhead light that was distracting to her. "We started to use candles when we were first living together," the woman told me, "because he wanted the lights on and I wanted them off—candlelight was the perfect compromise!"

The "lights on/lights off" dilemma is an important issue related to visual stimulation. In general, men are aroused by the sight of their partner's body and usually want some light in the room when they are making love. On the other hand, some women find that harsh light interferes with their experience of the other sensory stimulation. In addition, many women—and some men—are uncomfortable when their lover gazes at their body for fear that he or she will find it wanting in some way. Since many women are self-critical, they assume their lover is gazing at their body and having negative thoughts. This usually isn't true, of course, and sensitivity and open communication can usually make both partners relax and enjoy their bodies —at any age.

Lighting is also important when you "dress for sex." Erotic clothing includes lingerie, of course, but it also includes an array of practices such as donning favorite "costumes" and playing out fantasies. One couple I interviewed told me that undressing each other in front of a mirror arouses them, as does staying partially clothed while they have sex, with soft lighting in the room.

There isn't anything modern about using clothing as part of the sexual act. In the eighteenth century, Sarah, the first duchess of

Marlborough, wrote, "The Duke returned from the wars today and did pleasure me in his top boots."

An adventurous couple told me they enjoyed going out in public dressed in their sexiest clothing and then fantasizing about the responses of others around them. Away on a vacation one summer, they had what they termed "sex talk" in a restaurant one afternoon, and when they (quickly) returned to their hotel room, they fantasized about others in the restaurant stealthily following them and looking on as he removed her provocative lingerie before their lovemaking. The image of being followed and watched was an important part of the fantasy and satisfied an urge for exhibitionism, so it made sense that they'd dressed for the occasion!

Alex Comfort (author of *The New Joy of Sex*) calls sex the most important form of adult play, and fantasies and dressing for sex are part of the recreational value. Numerous books about adult sexuality emphasize the important role of fantasy and role playing, both of which involve visual stimulation. Sharing your fantasies and acting on them is a highly intimate activity, which is why many people shy away from this kind of sexual expression. But, as Dr. Comfort says, sex is one way adults can engage their imaginations and share a playful and private stage, much like they did as children.

The Sounds of Love

The sounds associated with lovemaking come in two basic varieties: the sounds we add, such as background music, and the sounds we make ourselves, both conversation and the sounds we make as we become increasingly aroused and reach orgasm.

In theory, anyway, adding music to eroticize your environment should be easy enough. Some individuals like the sensuous, sometimes frenzied beat of jazz or blues music; others like the soft tones of romantic instrumentals or classical music; still others prefer popular romantic ballads. Any of the above might be preferred at different times, depending on the mood of the lovers.

As with so many other sensory details, we tend to pay less attention as we move through the stages of a relationship. When was the last time you danced to romantic music in the living room? Or have you even asked your lover what kind of music he or she prefers during sex? Do you take a minute to turn on the radio or the tape player when you climb into bed and are thinking about sex? Most of the men and women I've interviewed commented that background music often makes them feel sensuous, but they seldom "bother" with it.

Adding music to the sexual environment is worth bothering with, because it is part of varying the sensory environment, and it helps block intrusive and distracting thoughts. It's also a shared sensory experience that can influence how you respond to your partner. Just as there is a rhythm to music, there is a rhythm to lovemaking, which can be influenced by the music in the background. A man who was very satisfied with his sex life told me that he and his wife are more likely to reach orgasm simultaneously when they have music playing because their bodies just naturally seem to synchronize as they move to the music. Music can intensify passion, and romantic lyrics express feelings that you may not have words for yourself. Vary the type of music you use and discover—or rediscover—the intimacy of a shared response to the backdrop of sound.

What Did You Say?

As any sex therapist will tell you, communication is the key to a good sex life. An important component of sexual communication takes place outside the bedroom. Most sex therapists agree that discussing our sexual needs or our fantasies when we're not actually engaged in sex is less threatening to both partners. If you feel the need to discuss some important sexual issues with your partner or a potential partner, then do so in a neutral environment. You'll both feel less vulnerable, and the conversation will be less stressful.

Another component of sexual communication takes place while we're making love. Some couples prefer to concentrate on the sensations of touch or scent, and talking becomes part of the afterplay. Other couples want to communicate intimately during sex, but seem to miscommunicate at least some of the time.

Do men really mean it when they say, "I love it when you talk dirty?" Well, yes and no. In the past, "ladies" weren't supposed to use crude or "naughty" language, so hearing a woman use sexual slang was breaking a taboo, which made it sexually exciting to some men. Men could use these terms in male company and with their lovers, but not in "mixed company." Once outside the bedroom, a woman was harshly judged if she used the same language that her lover wanted her to use in bed. To add to the confusion, not all men find the switch from "public nice" to "private naughty" particularly appealing.

Nowadays, the restrictions on women's language has changed somewhat, and the double standard has eased. But, most women are still careful about their language in many business and social settings because society's changes are uneven. In some business environments, a woman who freely uses sexual slang could be passed over when promotion time rolls around, and some men might interpret her language as a signal that she welcomes sexual advances from coworkers. Men are inhibited in their language, too, because they don't know if the particular woman they're dealing with will be offended.

All this confusion makes its way to the bedroom, and the only answer is mutual understanding and communication. Men need to understand that a woman's reluctance to turn on the "dirty talk" in bed may be because she finds the language offensive all the time. She can't easily turn her attitude off and on, and a sexual setting isn't necessarily going to change her feelings. Women need to understand that her partner isn't trying to demean her with sexual slang, which is how some women interpret a man's desire for this kind of erotic talk. The same words used on "the street" are heard as demeaning, but in the bedroom a man intends them to be earthy and arousing.

Mood is a factor, too. A woman I interviewed described it this way: "When I'm in a lusty mood, then talking dirty seems natural, and I like to hear the words and phrases myself. But when my mood is more romantic and tender, then the mood is broken if I sound like a character in a porno flick. I want my husband to tell me he loves me and talk in a low voice about how my skin feels or how I smell— romantic talk, not a crude inventory of my body parts. A playful mood calls for playful language, which may or may not be 'dirty.'"

This woman and her partner are able to match their sexual mood with their sexual language, an important sign of compatibility. Surveys have shown that the sexes are about even in their attitudes about "talking dirty" in bed; close to 60 percent of men and women find it at least acceptable.[1] This means that just over 40 percent of men and women don't find it appealing, so it's up to you to find out where your partner stands.

You're So Noisy!

Barry, a man I interviewed, said that he didn't pay that much attention to the words his lover used during sex, but the noises she made to show her pleasure fueled his passion. The moans and sighs and cries made during sex are generally involuntary, although some men and women suppress these noises because they don't want to be overheard.

Two University of California researchers conducted a study to determine if there is any particular significance to these "copulatory vocalizations," and they compared human couples to a pair of gibbons and members of a troop of chacma baboons.[2] They found that both men and women tend to sigh and moan during sex, and their respiratory rate changes, too. As women approach orgasm, their sounds intensify and have a regular, staccato "beat"; however, men's sounds are more irregular and don't have a predictable rhythm. Both sexes make explosive bursts of noises at orgasm, a fact I'm sure surprises no one.

The researchers found that gibbons and the chacma baboons, particularly the females, also made a variety of noises and calls, and these vocalizations had a complex pattern. They were most complex among the baboons, who change partners frequently. The vocalization pattern among the human couples and gibbons, both of whom tend to mate for longer periods, were less complex.

The two researchers believe that these vocalization patterns are too complex to be considered random or simply the by-product of the physical exertion involved in the sex act. Rather, they speculate that the calls have evolutionary significance and may play a part in encouraging primates to repeat copulation with the same mate. The cries also serve as a signal that one partner is reaching orgasm, which in human sexual affairs makes sense. Men and women tend to "read" the progression of sexual stimulation in part by the vocalizations of each partner.

It's also possible that the "sex call" is a way to let others in the group know that mating is taking place, a more primitive way of ensuring some privacy. In the higher primates, the sounds were also a way to keep other males away from the female. Our primate relatives are uninhibited about their sexual moans and cries, whereas we have tended to evaluate them and, in some cases, have attempted to control these essentially involuntary sounds, built into our sexual physiology. Because humans have language, we also add words and phrases to our bedroom talk and our utterances at the point of orgasm. These vocalizations intensify eroticism and are integral to the sexual experience.

While you don't want to wake the children or the neighbors, neither should you attempt to suppress these expressions of passion. Crying or laughing at the point of orgasm is also common and appears to be related to emotional release. A woman I interviewed told me that her partner became upset when the intensity of her orgasm caused her to cry because he thought she was upset or sad, but when she laughed he was relieved. Her crying, however, was closely related to her response to the joy of the intimacy, not to sadness. The best policy is never to judge your partner's reactions. Laughing, crying, sighing—whatever—it's all part of the wide range of emotional responses to sex.

It's interesting to note that the sounds we make during the arousal stage of lovemaking are similar to the sounds we make in response to an appealing odor. You may lift the lid of a simmering stew and exclaim "ahh" or "ooh," and that is the same sound you make when your lover strokes an erotically sensitive part of your body.

Eros and the Bedroom

Obviously, most couples aren't so rigid that everything has to be perfectly arranged for lovemaking. The spontaneity of the moment is lost if you and your partner are pleasantly aroused but you must first alter the lighting, find the scented candles, and change your lingerie. You probably don't have time for a sensuous massage every time you have sex with your lover. The time and place may not always be perfect, and the sex can't scale the heights of passion during each encounter, either.

What you can do is begin paying attention to the erotic environment you're creating and devote some time to this important element of your relationship. Evaluating the atmosphere in your bedroom is a good place to start.

Regardless of the allure of exotic locations for making love, most couples have the majority of their sexual encounters in their bedrooms.

But the function of this space has changed over time. Decades ago, bedrooms in most homes and apartments were designed as fairly small rooms, big enough for a double bed, perhaps a nightstand or two, a chest of drawers and a dresser, but little else. A couple's bedroom was a private place, where the activities centered around the bed—sex and sleep and perhaps some mystery novels on the nightstand. But as lifestyles changed, so did our bedrooms.

Nowadays, architects and realtors refer to "master bedrooms," and in new homes, this room is often big enough for a television and VCR, a treadmill or exercise bike, a computer table, and perhaps even some file cabinets. Look around your own bedroom. How sensual is it? What kind of atmosphere have you created? Is it really a relaxing, intimate retreat, a place set aside for restful sleep and lovemaking? The beds we typically sleep in are bigger than ever, but they may not dominate the room we call the bedroom. All the paraphernalia we put in these large rooms have *de*-eroticized the very room we most associate with erotic life. Nowadays, we might even take trips to cyberspace through our on-line services as we're lounging around in bed, supposedly relaxing!

When couples visit sex therapists, they are often asked to describe their bedrooms and the variety of activities that take place there. A man I interviewed had sought treatment for diminished sexual desire from a sexual dysfunction clinic, and the kinds of questions he and his wife were asked puzzled him at first. But over time, he began to see that the atmosphere in the bedroom was contributing to the problems that were creating difficulties in his marriage. Does his description sound familiar?

When the therapist first asked about our bedroom, I dismissed the questions as silly. What difference does it make if we have a television in the room? Having it there doesn't mean we have to watch it, but of course we did. We fell asleep with the television on almost every night. We both brought our work to bed, too. There we were, typical nineties professionals with our laptops and files spread around us—in our bed! And of course our bed is huge—the biggest we could find. What should have been an erotic playground had turned into a workstation—an extension of our offices.

One night, my wife and I began using our calculators to figure out if we could afford a trip to Europe, and I began to complain about a couple of things she had purchased for the house. Her feelings were hurt by my implied criticism, and before long, we were arguing about money. I began to think about my parents' bedroom, which was a small, cramped room. That's how I saw it as a child, but now I look at it differently. They watched television in the living room and paid their bills and worked out their budget at the kitchen table. I recall that their bedroom was a restful place—their private room. When they closed the bedroom door, there was nothing to distract them from each other. My wife and I watched sex on television all the time, but we rarely made love with each other. That seems bizarre now that we've changed our lives.

As part of their therapy, this couple was asked to remove all the distractions from their bedroom, including the television. If they were bringing work home from the office, then they were to do it in another room—no computers and calculators in the bedroom. At first, they went through "withdrawal." They had to change their routine, and that was tough, because like all of us, they were creatures of habit. Until it was brought to their attention, they saw nothing unusual about watching television and writing reports in bed. But over time, they changed the focus of their room, and they altered their lifestyle as well. As a result, their sex life improved, and both partners were much happier.

I realize that you may live in a small apartment or a home that is already overflowing with "necessities." This is not a book about simplifying your life; an abundance of valuable material is available about evaluating your possessions and lifestyle. But I am urging you to take the advice of sex therapists and rethink the function of your bedroom. Must it truly double as a health club, an office, or an entertainment center? (For a few months at a time, your bedroom may serve as your baby's first nursery, and that's fine. It's convenient and reassuring to keep small babies near their parents.)

Studies and surveys have determined that Americans are a tired society in part because we cheat ourselves of much-needed sleep when we watch television late at night and then drag ourselves out of bed in the morning. Change the arrangement of your home if you must, but think about returning to the concept of the bedroom as center of emotional intimacy and rest. Leave all the mundane stimulation outside the bedroom door. You'll find that you and your partner will have a chance to just be together without the common and numerous distractions. Watch a movie together—even an erotic one—in some other room. By all means, discuss your finances and family problems away from the bedroom. If you want to play video games or do research on the Internet, move the computer out of the bedroom.

Imagine an ideal bedroom and think about its sensual and romantic atmosphere. What can you change in your room right now? Can you decorate the room with plants and fresh flowers? Can you vary the lighting with three-way lightbulbs or dimmer switches? Is there a convenient way to play your favorite sensuous music without the distraction of blaring commercials? (Listening to the clock radio blast news and traffic reports before you're even fully awake is not necessarily the best way to start your day, either.) Have you arranged and decorated the room in such a way that it *feels* like a private and intimate retreat? Does it smell appealing when you open the door?

Can you begin adding sensuous and relaxing odors to contrast the mood in this space with the rest of the house?

There has never been a better time to begin adding odors to your home environment. And in the next five to ten years, odors will become an even more significant part of interior decorating. We already have an incredible range of choices of colors, fabric, and lighting. A flick of a switch can bring music into your bedroom. You can use the sensory tools available to nurture yourself and your lover and create an environment that will make your home, including the bedroom, the haven it was meant to be.

Remembering the Sexy Nose

I hope this book has helped you examine your attitudes and feelings about your sexual relationships. The emphasis has been on odors, and I believe that you can enhance the pleasures of sexuality by understanding your powerful and sexy sense of smell. By using a variety of scents in your environment and on your body, you are "feeding" your nose what it needs to help you enjoy your sexuality to its fullest. Your nose is one of your sexiest organs, so give it the attention it needs, and enjoy the results!

Chapter Notes

CHAPTER 1

[1] H. Sugano, "Effects of Odors on Mental Function" (abstract), *JASTS* 22 (1988): 303.

[2] A. R. Hirsch and L. H. Johnson, "Odors and Learning," *Journal of Neurological and Orthopedic Medical Surgery* 17 (1996): 119–26.

[3] H. Ellis, *Sexual Selection in Man* (F. A. Davis Co., 1906), 65.

[4] E. Frank, C. Anderson, and D. Rubenstein, "Frequency of Sexual Dysfunction in 'Normal' Couples," *New England Journal of Medicine* 299 (July 20, 1978): 111.

CHAPTER 2

[1] W. Velle, "Sex Differences in Sensory Functions," *Perspectives in Biology and Medicine* 30, no. 4 (summer 1987): 491–522.

[2] R. H. Porter et al., "Recognition of Kin through Characteristic Body Odors," *Chemical Senses* 11, no. 3 (1986): 389–95.

[3] R. L. Doty et al., "Sex Differences in Odor Identification Ability: A Crosscultural Analysis," *Neurophysiologia* 23 (1985): 667–72.

CHAPTER 3

[1] M. N. McClintock "Menstrual Synchrony and Suppression," *Nature* 299 (1971): 244–45.

[2] G. Preti et al., "Determination of Ovulation and Alteration of Menstrual Cycle by Human Odors" (abstract), 188th National Meeting, Amer. Chem. Soc., Philadelphia, (August 1984).

[3] C. A. Graham and W. C. McGrew, "Menstrual Synchrony in Female Undergraduates Living on a Coeducational Campus," *Psychoneuroendocrinology* 5 (1980): 245–52.

[4] J. Durden-Smith and D. DeSimone, *Sex and the Brain* (New York: Arbor House, 1983), 215.

[5] G. Preti et al.

6 R. R. Gustavson, M. E. Dawson, and D. G. Bonnett, "Androstenol, a Putative Human Pheromone, Affects Human (Homo Sapiens) Male Choice Performance," *Journal of Comparative Psychology* 101 (1987): 210–12.

7 M. Kirk-Smith et al., "Effect of Androsterone on Choice of Location in Others' Presence," in *Olfaction and Taste* VII (Oxford: IRL Press, 1980), 397–400.

8 M. Kirk-Smith et al., "Human Social Attitudes Affected by Androstenol," *Research Communication in Psychological Psychiatry and Behavior* 3 (1978): 379–84.

9 Ibid.

10 N. M. Griffiths and R. L. S. Patterson, "Human Olfactory Responses to 5 a-androst-16-en-3-one—Principal Component of Boar Taint," *J. Agric. Fd. Chem.* 21 (1970): 4–6.

11 N. M. Morris and J. R. Urdry, "Pheromonal Influences on Human Sexual Behavior: An Experimental Search," *Journal of Biological Science* 10 (1978): 147–57.

12 K. Larsson, "Impaired Mating Performances in Male Rats after Induced Anosmia Peripherally or Centrally," *Brain, Behavior, and Evolution* 4 (1971): 463–71.

13 R. L. Doty et al., "Endocrine, Cardiovascular, and Psychological Correlates of Olfactory Sensitivity Changes during the Human Menstrual Cycle," *Journal of Comparative Physiological Psychology* 95 (1981): 45–60.

14 M. D. Kirk-Smith, S. Van Toller, and G. H. Dodd, "Unconscious Odour Conditioning in Human Subjects," *Biological Psychology* 17 (1983): 221–31.

CHAPTER 5

1 R. H. Porter and J. D. Moore, "Human Kin Recognition by Olfactory Cues," *Physio. Behavior* 21 (1981): 493–95; and R. H. Porter et al., "Maternal Recognition of Neonates through Olfactory Cues," *Physio. Behavior* 30 (1983): 151–54.

2 R. H. Porter et al., "Recognition of Kin through Characteristic Body Odors," *Chemical Senses* 11, no. 3 (1986): 389–95.

3 T. Lord and M. Kasprazak, "Identification of Self through Olfaction," *Perceptual and Motor Skills* 69 (1989): 219–24.

4 M. J. Russell, T. Mendelson, and H. V. S. Peeke, "Mothers' Identification of Their Infants' Odors," (1982, in preparation).

5 W. S. Cain, "Odor Identification by Males and Females: Predictions vs. Performance," *Chemical Senses* 7 (1982): 129–42.

6 R. L. Doty et al., "Communication of Gender from Human Axillary Odors: Relationship to Perceived Intensity and Hedonicity," *Behavioral Biology* 23 (1978): 373–80.

7 Z. T. Halpin, "Individual Odors among Mammals," in *Advances in the Study of Behavior* 16 (1986): 51.

8 A. R. Hirsch and G. S. Bussell, "The Effects of Inebriation on Olfaction," *Journal of Investigative Medicine* 43 (September 1995): 422A.

CHAPTER 6

1 K. Larsson, "Impaired Mating Performance in Male Rats after Anosmia Induced Peripherally or Centrally," *Brain, Behavior, and Evolution* 4 (1971): 463–71.

2 R. L. Moss, "Modification of Copulatory Behavior in the Female Rat Following Olfactory Bulb Removal," *Journal of Comparative and Physiological Psychology* 74 (1971): 374–82.

3 A. R. Hirsch, "Olfaction and Anxiety," *The Clinical Psychiatry Quarterly* 16, no. 1 (1993): 4.

4 H. Sugano, "Effects of Odors on Mental Function" (abstract), *Japanese Association for the Study of Taste and Smell* 22 (1988): 2.

5 P. D. Maclean and D. W. Ploog, "Cerebral Representation of Penile Erection," *Journal of Neurophysiology* 25 (1962): 29–55.

CHAPTER 7

1 A. C. Kinsey et al., *Sexual Behavior in the Human Female* (Philadelphia: W. B. Saunders, 1953).

2 E. Chesser, *The Sexual Marital and Family Relationships of the English Woman* (London: Hutchinson's Medical Publications, 1956).

3 E. Frank, C. Anderson, and D. Rubinstein, "Frequency of Sexual Dysfunction in 'Normal' Couples," *New England Journal of Medicine* 299 (July 20, 1978): 111.

4 R. C. Kolodny, W. H. Masters, and V. E. Johnson, *Textbook of Sexual Medicine* (Boston: Little, Brown and Company, 1979): 321–51.

5 J. J. Greer, P. Morokoff, and P. Greenwood, "Sexual Arousal in Women: The Development of a Measurement Device for Vaginal Blood Volume," *Archives of Sexual Behavior* 3 (1974): 559–64.

6 A. R. Hirsch, "Odors and the Perception of Room Size," presented at 148th Annual Meeting of the American Psychiatric Association, Miami, 1995.

7 A. R. Hirsch and C. Kang, "The Effects of Green Apple Fragrance on Migraine Headache," *Headache* 37, no. 5 (1997): 312.

CHAPTER 8

1 A. R. Hirsch and T. J. Trannel, "Chemosensory Dysfunction and Psychiatric Diagnoses," *Journal of Neurological and Orthopedic Medical Surgery* 17 (1996): 25–30.

2 As cited in B. Chichester, K. Robinson, and the editors of Men's Health Books, *Sex Secrets: Ways to Satisfy Your Partner Every Time* (Emmaus, Pa.: Rodale, 1996), 13.

CHAPTER 9

[1] J. Mensing and C. Beck, "Fragrance from a Psychological View: The End of a Myth," in *H & R Book of Perfume* (London: Johnson, 1984).

CHAPTER 10

[1] J. R. King, "Anxiety Reduction Using Fragrances," in *Psychology and Biology of Fragrance* (London: Chapman and Hall, 1988), 147–65.

CHAPTER 11

[1] E. Frank, C. Anderson, and D. Rubinstein, "Frequency of Sexual Dysfunction in 'Normal' Couples," *New England Journal of Medicine* 299 (July 20, 1978): 111.

[2] As cited in B. Chichester, K. Robinson, and the editors of Men's Health Books, *Sex Secrets: Ways to Satisfy Your Partner Every Time* (Emmaus, Pa.: Rodale Press, 1997), 34–35.

CHAPTER 12

[1] S. Bechtel, L. R. Stains, and the editors of Men's Health Books, *Sex: A Man's Guide* (Emmaus, Pa.: Rodale, 1996), 279.

[2] W. Hamilton and P. Arrowood, "Copulatory Vocalizations of the Chacma Baboons, Gibbons, and Humans," *Science* 200, no. 4348 (June 23, 1978): 1405–9.